SAVE DOLLARS AND BEAT INFLATION WITHOUT SACRIFICE!

- 15 Ways to Save on Heating and Cooling
- 7 Ways to Save on Your Car
- 6 Ways to Save on Hot Water
- 4 Ways to Save on Cooking
- 5 Ways to Save on Lighting
- 4 Ways to Save on Woodburning
- 8 Ways to Save Energy and Money on Necessities

PLUS—Little Ways to Save a Lot, and a Plan of Action So You Can Begin Saving Now!

NO-COST LOW-COST ENERGY TIPS

52 Ways to Save $1,000 a Year
in Energy Costs
Without Sacrifice

by
Stuart Diamond

BANTAM BOOKS
TORONTO · NEW YORK · LONDON · SYDNEY

NO-COST LOW-COST ENERGY TIPS
A Bantam Book / September 1980
2nd printing .... November 1981

ISBN 0-553-14239-9

Published simultaneously in the United States and Canada

Bantam Books are published by Bantam Books, Inc. Its trademark, consisting of the words "Bantam Books" and the portrayal of a rooster, is Registered in U.S. Patent and Trademark Office and in other countries. Marca Registrada. Bantam Books, Inc., 666 Fifth Avenue, New York, New York 10103.

PRINTED IN THE UNITED STATES OF AMERICA

11 10 9 8 7 6 5 4 3

Contents

8. Utility Companies and Banks

9. Buying and Building New Things

10. A Plan of Action

Appendices

No Sacrifice!

By now we know the bad news. All we have to do is look at our energy bills. We've also heard the proposed remedy: Sacrifice.

To this I say, bunk!

Sacrifice is unnecessary. Moreover, we didn't come all this way to go back to a Spartan life. We are not interested in freezing in the winter and roasting in the summer. We didn't buy a car to watch it sit in the driveway. *And we don't believe the answer is paying more.*

For these reasons, I wrote this book. I've interviewed more than 5,000 people in the past few years, and discovered that most of the quick, cheap, easy answers don't involve sacrifice. In fact, just the opposite. A few, quick, energy conservation measures will save money and *improve* our life-style, not make it worse. Clearly, $50 saved in energy costs might mean dinner at a fancy restaurant.

I also looked around and found books that list 537 ways to save energy, without telling how, how much it will cost, and how much it will save. None of us wants to do 500 things. I'd like to do maybe 10 things, fast, and then get on with the rest of my life.

So I've included only the best, most cost-effective things I could find. There are also few maybes—clearly marked as such—because you've heard about them. I wrote as few words as I could: There's already too much information.

You will be able to do many of things in this book in a weekend, saving perhaps hundreds of dollars. But, in general, don't overload yourself. Do only what's comfortable. Do a few things at a time—even one thing at a time. See what happens. Then perhaps try something else.

You will have to become comfortable with a few everyday items. One is the phone book. The Yellow Pages list contractors of all types. The White Pages list local, state and federal agencies of all kinds. In each case I've listed where you should look, giving addresses where appropriate.

If you are concerned about the safety of any device, ask for certification on the label by the federal government or a nationally recognized testing laboratory such as Underwriter's Lab. UL has local offices around the country.

A word for renters. If you pay directly for utilities, many of the measures in this book will be economical for you, since they cost only a few dollars. For more expensive items, you might share the cost in some way with the landlord. If you don't pay directly for utilities, it means your landlord is passing the cost directly to you—and probably has little incentive to reduce the cost. Try to pay for utilities directly. Or, give this book to the landlord, so he can reduce his costs and, ultimately, yours.

All the estimates in this book are approximate, to be used for comparison purposes. If you added them up, the fuel company would owe you money—so clearly, there's overlap. But the bottom line is that many of you can save more than half—perhaps two-thirds—of your energy costs with a modest investment.

My own home can serve as an example. I bought an old Victorian house in late 1979: a "workman's special" that guzzled fuel to heat its five bedrooms. The first thing I did was an energy audit with my index finger. I went all around the house—under doors, around windows—and everywhere I felt a draft, I put caulk or weatherstripping. I tuned the furnace, and did a few other things outlined in this book.

Result: My fuel bill for the winter was *less* than the previous year, even though the price of heating oil had doubled. And I kept my thermostat at 70 degrees when I was home.

There is no magic to winning the energy war. Victory comes in small fixes. A little here, a little there. A dollar here, a dollar there. A journey of a thousand dollars begins with a single tube of caulk.

Acknowledgments

The facts and figures I quote come from a multitude of sources collected for several years. But I would specifically like to mention a few sources whose information has contributed much to my energy understanding. They include the National Bureau of Standards, various technical experts at the U.S. Energy Department, the consumer offices of many utilities, and numerous consumers that sent information and then spent hours with me on the phone.

Finally I wish to thank two people: Paul S. Lorris, director of municipal energy management for Boston. He understands energy as few people do. And Kim Greer Verzyl, whose insights any writer would cherish.

I also wish to acknowledge the U.S. Energy Department's pilot program on low-cost, no-cost energy ideas. We share the same philosophy, although a greater federal commitment is needed. Hopefully, this book will aid the government's effort.

1
Heating and Cooling

Tip #1
Children, Dogs and Doors

Description: Open doors cost money. Even for 30 seconds, they allow valuable heat to rush out during the winter and rush in during the summer.

— Leaving the door ajar while taking out the garbage or going to the car is like taking a quarter and flipping it into the gutter. Do it 200 times a year, it's $50 down the drain.

— Each child or pet can increase your heating and cooling bills 5 percent, by going in and out a lot and not closing the doors properly. Thus, two girls, a boy and a cocker spaniel can raise your heating and cooling bills 20 percent.

Remedy: Train your children. Train your pets. Train yourself. Do not have last-minute conversations with departing guests in an open doorway. It will make the evening much more expensive.

Cost: $0.

Annual Savings: $50 or more. Five year savings, $300 or more.

Time to install: None.

Epilogue: There are more than 500 million doors in the U.S. Without doubt, they are left open unnecessarily for more than 5 billion minutes a year. Cost: more than $1 billion—money that would pay for a lot of jobs.

What's wrong with this picture?

Tip #2
Fooling the Thermostat

Description: A clever, cheap clock thermostat that saves money without sacrifice. You can install it yourself.

— It adjusts your thermostat when you are asleep or at work. You don't notice the difference, but your pocketbook does.

— In winter, each 8-hour, 1 degree set-back cuts heating bills 1 percent. A 10-degree, night set-back saves 10 percent. In summer, each 1 degree, 8-hour thermostat increase cuts cooling costs 2 percent.

— Most clock thermostats must be installed by an electrician and cost $60–$100 or more.

Remedy: A "thermostat fooler," a small 20-watt heating coil you affix to the wall about 3 inches under your normal thermostat. The coil is attached to a timer that plugs into a conventional wall outlet.

— In winter, set the timer for an hour after you go to sleep or work. The timer will activate the little electric coil, which warms the air under the thermostat. The thermostat thinks the room is warm, so it doesn't turn on the heating system. The house, meanwhile, cools down.

— An hour before you arise or return, the coil shuts off, the thermostat senses the real room temperature and turns on the heating system. The fooler works in reverse for air conditioning.

Cost: $20 to buy; $5 in annual electricity.

Savings: $100–$200 in first year; $1,000 after 5 years.

Time to Implement: 15 minutes.

Where to buy: Hardware stores, home improvement outlets. Or order one that was developed under the U.S. government's Energy Related Inventions Program: Flair Manufacturing Corp., 600 Old Willetts Path, Hauppauge, N.Y. 11787.

A "thermostat fooler" that you install yourself.

Tip #3
Plugging Basketballs, I

Description: If you added up all the cracks in an average house and put them together, they would form a hole in the wall the size of a basketball. All winter long, heat leaks out. All summer long, heat leaks in.

Remedy: Caulking and weatherstripping, invented in ancient times and all-but-forgotten in modern homes.

— Caulk is a putty-like substance that comes in a tube. It's a sealer. You squirt it in small cracks around window frames, chimneys and other areas where objects go through your walls and ceilings.

— Weatherstripping is a felt, metal, foam or plastic strip that you put around doors and on the moving parts of windows. Plug the cracks under front and back doors first. If you feel drafts through windows when the wind is blowing, you need weatherstripping.

Cost: $40–$50 for a whole house.

Savings: $100 a year in heating bills for a leaky house, $50 in cooling costs. Five year savings, $500 or more.

Where to buy it: Any hardware store. There are many different types. Discuss it with the hardware clerk; compare price and longevity. Lasts 3-20 years.

Epilogue: Less than a third of U.S. homes are properly caulked and weatherstripped. That's a lot of basketballs.

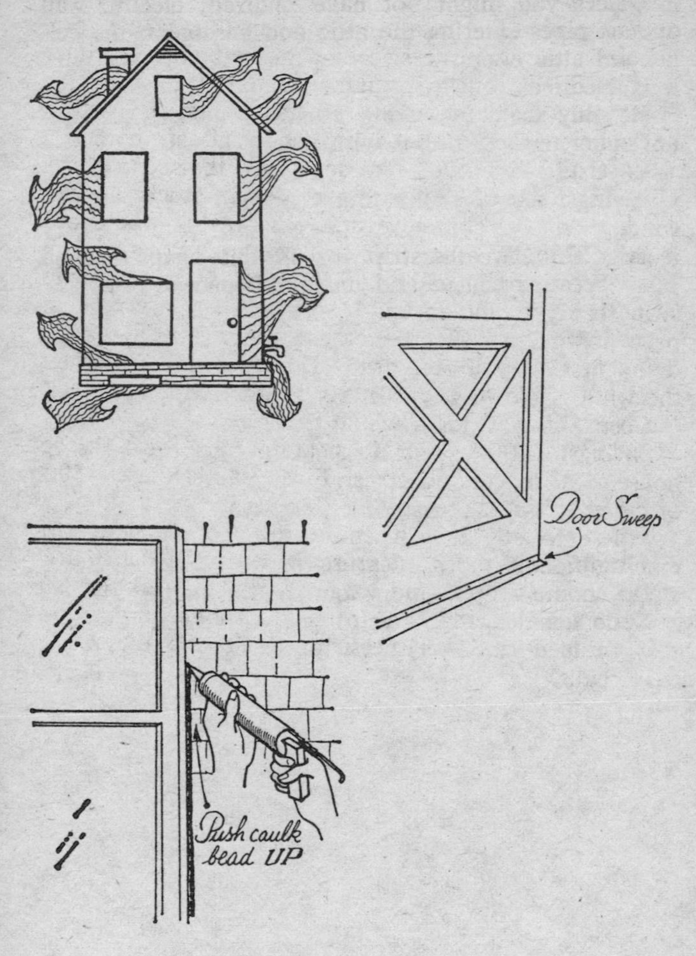

Why suffer? Plug that draft.

Tip #4
Plugging Basketballs, II

Description: A lot of money leaks from your house in places you might not have realized: electric wall outlets; pipes entering the attic and basement; the cellar and attic doors.

Remedies:

(1) Buy foam insulating gaskets for electrical outlets. Plug unused outlets with plastic inserts.

Cost: $10.

Savings: $10–$20 the first year, $60–$125 after 5 years.

(2) Caulk, weatherstrip, or insulate gaps around pipes between heated and unheated spaces.

Cost: $2.

Savings: $25–$100 the first year, $150–$750 after 5 years.

(3) Insulate and weatherstrip the back of doors to unheated interior spaces, such as attic, cellar and crawl spaces.

Cost: $2–$5.

Savings: $8–$20 the first year, $50–$125 after 5 years.

Where to buy it: Local hardware stores, lumberyards, home improvement outlets. Insulating gaskets may be somewhat difficult to find. Call first.

Time to implement: A few hours.

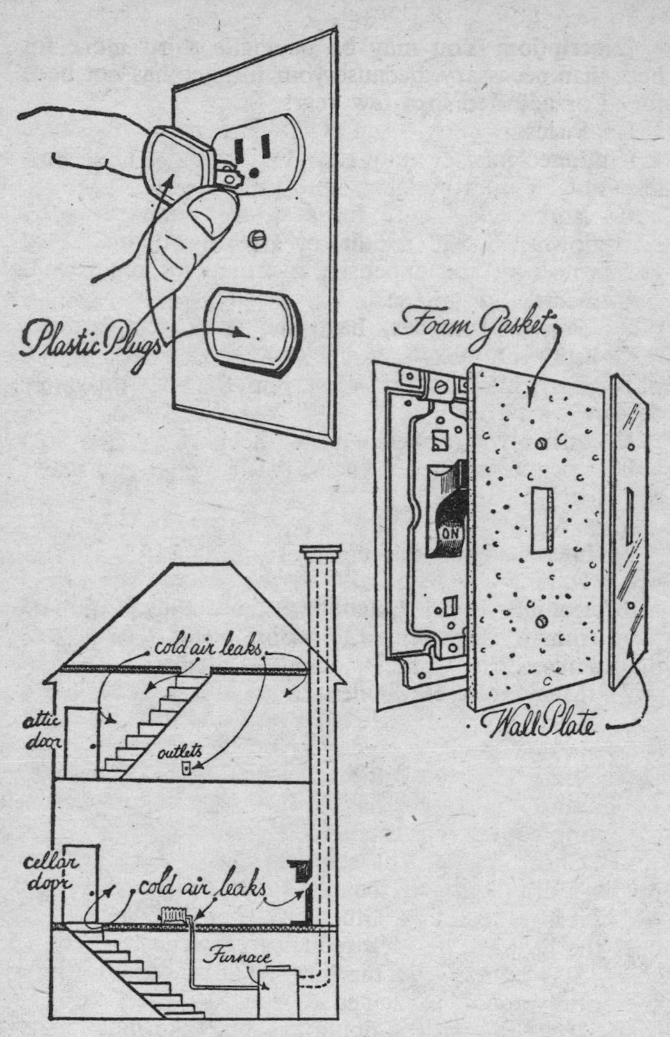

Spend $14 once, save more than $100 per year.

Tip #5
Out-of-Tune Furnaces

Description: You may be paying a third more for heat than necessary, because your furnace has not been tuned or adjusted in a few years.

— Furnaces should run at 70–80 percent efficiency. An untuned furnace may run at 50 percent efficiency, or less.

— Improperly tuned furnaces are more prone to breakdowns in the middle of freezing nights. That means no heat and expensive overtime for repairmen.

Remedies:

— For oil furnaces, have the stack temperature, combustion products and efficiency checked. If it's not part of your fuel service contract, call a heating contractor.

— For gas furnaces, adjust the air/fuel mixture (blue flame instead of yellow); make sure the jets are not blocked. Call your utility company or a gas appliance repairmen found in the Yellow Pages.

Additional advice:

— A tune-up should also include: Removal of soot and cleaning of heat exchanger coils, which transfer heat from the flame to the air or water that warms your house.

— Many hot water boilers have baffles: metal plates that sit among the heat exchanger coils. They radiate heat back to the coils and slow down hot combustion gases. Both improve boiler efficiency up to 5 percent. Make sure your boiler has all its baffles. If not, have a heating contractor install them.

— Check the oil line solenoid valve to make sure oil isn't dripping into the combustion chamber when the unit is not running. Otherwise, soot and smoke will coat the heat exchangers, reducing efficiency.

— Stay home when the furnace is tuned, and have the entire process explained to you.

— If a contractor is doing the job, take bids.

— If your furnace is below 70 percent efficiency after it's tuned, get a new burner. (See Tip #52.)

Cost: $25 or so for a tune-up. $2–$50 for baffles.

Annual savings: Up to $300 for tune-up, $25–$50 for baffles.

Time to implement: An hour or so.

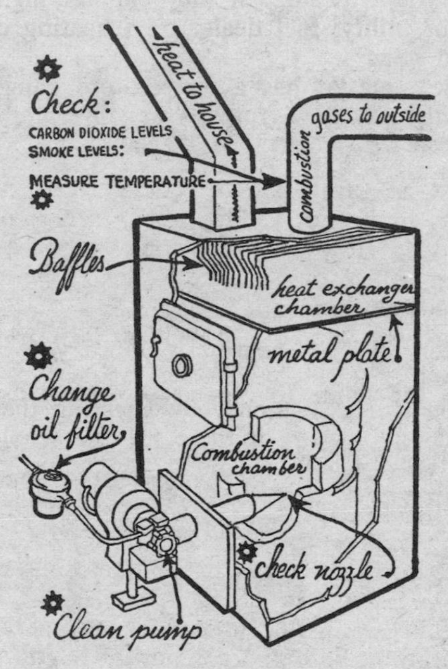

Don't throw away hard-earned money on an untuned furnace.

Tip #6
Too Hot Heat

Description: In many hot water heating systems, the furnace heats the radiator water to 200 degrees, when only 70-degree heat is needed in the house. Result: Fuel waste.

Remedy: Check the "aquastat," which governs the water temperature in your heating system and is located on or near the furnace. DON'T DO IT YOURSELF if you're unsure: It may contain high voltage. Have your utility, fuel dealer or a heating contractor show you how.

— The aquastat has a temperature gauge, with a water *outlet* from the furnace and a water *inlet*—or return—from your rooms. The outlet should always be 20 degrees hotter than the inlet. Usually, 160/outlet and 140/inlet are adequate settings.

— The trick is to set the aquastat at the lowest temperature that keeps your house warm. The colder the outside temperature, the higher the aquastat temperature needs to be. On the other hand, the more you tighten up your house, the more you can lower the aquastat.

— You will save money by lowering the aquastat somewhat and leaving it there for the whole season. But you will save more money by adjusting it 2–3 times during the season: lowering the aquastat temperature more in the fall and spring, and less in the heart of winter.

Cost: $0–$20. Should be part of most fuel service contracts.

Savings: Up to $200 the first year, $1,250 after 5 years.

Time to implement: 2 minutes.

Epilogue: For $275 or so, you can buy an aquastat that automatically and continuously adjusts itself, according to the temperature of the outside air. This would always keep the aquastat at the lowest temperature needed to heat your house.

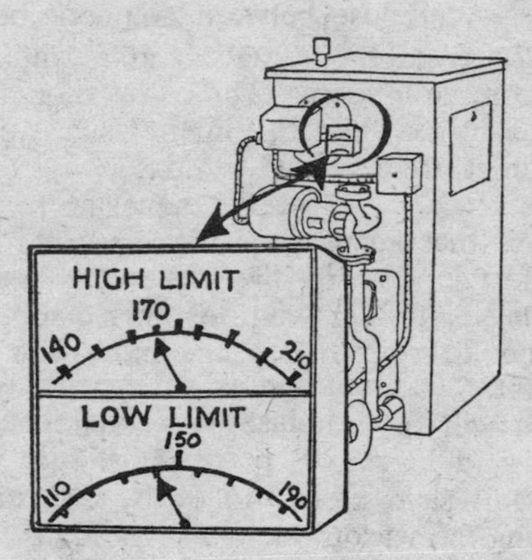

$200 saved for 2 minutes' work.

Tip #7
Insulation: Roof Yes, Walls Maybe

Description: If you don't cap your house with insulation, the warmth you paid for just rises through your roof, heating the sky. Even today, only one in five homes has enough insulation.

— Wall insulation, advertised widely, is one of the least economical things you can do in an existing house. In new construction it's great. In existing homes it's often done wrong. There are many pitfalls.

Remedies:

— At least 6 inches of attic insulation. Lay fiberglass or pour cellulose between the floor beams, with the aluminum foil vapor barrier between the insulation and the heated area.

— You can take up a finished attic floor in one day with a nail-removing tool called a cat's paw, sold in hardware stores.

— With a contractor, get a lot of estimates. If you want a heated attic, put fiberglass between exposed rafters or have cellulose blown in. Follow guidelines listed below.

— Since wall insulation is so expensive ($750 or more), it should be done last—after the cheaper things. The insulation's 10–20 percent savings may then amount to only $50–$75 a year. Potential health problems of blown-in urea formaldehyde are being investigated by the federal government.

— Blown-in insulation may settle or otherwise not cover all interior wall spaces. If you do it, however: (a) get a guarantee all walls will be filled; check it with an infrared (heat-sensitive) picture, and (b) make the contractor guarantee to pay for any health or fire problems.

Where to find it: Lumberyards, Yellow Pages.

Cost: Attic insulation, $200–$400.

Savings: 30 percent of heating bills (perhaps $300 per year); 20 percent of cooling bills.

Measure attic. Multiply square footage by .9 if 16-inch studs, by .94 for 24-inch studs, to get insulation area.

Tip #8
Insulation You Forgot

Description: Improperly sealed pipes and ducts in the cellar, attic and crawl spaces can waste a lot of money.

— A study by Oak Ridge National Laboratory found that leaking hot air ducts in unheated places raised fuel bills up to 25 percent. Uninsulated air conditioning ducts through hot attics send cooling bills through the roof.

Remedy:

— Run your finger along heating and air conditioning ducts to find leaks. Seal with duct tape sold in hardware stores. Then wrap 3½ inch thick fiberglass insulation around the ducts, with the silver vapor barrier on the outside.

— Also wrap fiberglass insulation around hot water pipes. Home improvement outlets sell special thin duct insulation, but it costs several times as much as fiberglass batts sold at the same store or at a lumberyard. Cut the insulation to size.

Cost: $10.

Savings: Up to $125 the first year; $750 after 5 years.

Time to install: An hour or two.

Epilogue: After you finish this job, consider insulation on the ceiling of an unheated cellar: It will make the floor of your living area above much warmer. Hold the insulation to the ceiling with chicken wire. The job may cost $200, but will save perhaps $100 per year. Always install the aluminum foil vapor barrier toward the heated area.

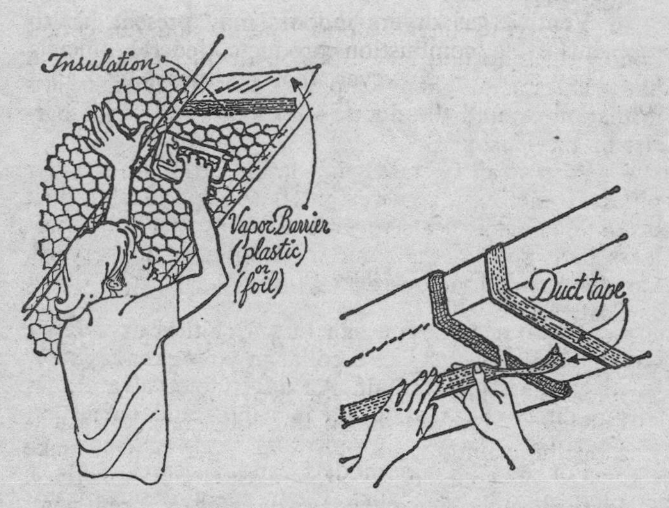

Big savings on insulation you forgot.

Tip #9
Heat from the Clothes Dryer

Description: Many clothes dryers send $50 worth of heat outside through the exhaust vent each winter.

— In addition to losing the dryer's heat, you may also be losing expensive air warmed by your home's heating system.

Remedy: Vent electric dryers indoors during the winter. You will not only save money in heat: You will also add moisture, reducing the need to humidify the dry indoor air.

— Venting gas dryers indoors may present health problems from combustion products, federal officials say. They concede, however, that gas ranges also emit combustion products into the kitchen without hazard. No definitive studies have been done. The matter is very much up in the air.

— To do it yourself, remove the exhaust vent from your wall or window and plug the hole with insulation or with rags stuffed in a plastic bag. Then, remove the vent hose from the machine or drape the hose into a garbage can.

— To be neater, you can buy a clothes dryer heat reclaimer. It attaches like a collar to your exhaust hose, which you first cut in half. An adjustable valve allows you to direct the heat inside in winter and outside in summer.

— For home-made models, add cheesecloth or a lint screen to the exhaust. Clean the lint screen frequently on all models.

Cost: $0–$10.

Savings: $20–$50 the first year; $150–$350 after 5 years.

Time to install: 30 minutes.

Where to buy: Local home improvement centers, or from:

(1) Bede Industries, 1985 W. 85th St., Cleveland, Ohio.

(2) In-O-Vent Energy Savings Products, 7295 Cascade Woods Drive, Grand Rapids, Michigan.

A small investment saves money and moisture in the winter.

Tip #10
Plastic Storm Windows

Description: If you can't afford glass storm windows, you can still cut heating or cooling bills 15 percent by adding a layer of plastic to your single-pane windows.

— The cheaper plastics will not give you a very good outside view, but often you don't need one: on cellar or attic windows, and windows always covered with drapes.

— The fuel savings with plastic storms may give you enough money to eventually buy glass storms.

Method: Buy a roll of 4 mil (4/1000th-inch thick) polyethylene plastic sheeting at a hardware store.

— First measure the width of your largest window so you know how wide a roll to buy. Measure the length of all windows so you know how long a roll to buy.

— Attach the plastic to your windows by duct tape, clear weatherstripping tape, staples or thumb tacks. Leave a 1-inch air space between the plastic and your single-pane window.

— You can also wrap the plastic around cardboard or wooden slats and attach the slats to the window frame.

More expensive models:

— Plastic storm window kits with tacks and slats are more convenient but cost three times the method described above.

— Clear, high-quality plastic and transluscent metal films are generally not economical for homes. Use the cheaper plastic and save for glass storms.

Cost: $15 for a whole house with polyethylene rolls.

Savings: Up to $150 per year.

Time to install: 2–3 hours.

Cheap answer to storm windows. Saves up to $150 per year.

Tip #11
Diet for a Furnace

Description: Most oil and gas furnaces in this country are too large and powerful for the homes in which they are located. As a result, they waste fuel.

— Because furnaces are "oversized," they cycle on and off. They start up, inject a burst of heat into your heating system, and then shut off.

— Machines that start and stop a lot are quite inefficient. Consider cars in stop-and-go traffic. An oversized furnace can increase heating bills by 25 percent.

— As you make your house more energy conserving by adding insulation, caulking and other measures, your furnace will become even more oversized and inefficient, because fewer demands will be placed on it.

Remedy: A smaller fuel nozzle, which permits less fuel into the furnace at any time. This makes the furnace run longer to meet the house's needs. The burner cycles less and becomes more efficient.

— This is called "underfiring." Different heating systems have different minimum nozzle sizes. In no case should the nozzle's output be less than half a gallon per hour. Your contractor should know the correct size. Also, have a proper oil filter installed to remove grit that might clog a smaller nozzle.

Cost: $5 plus installation.

First-year savings: Up to 8 percent of your fuel bills. $80 the first year; $500 after 5 years.

Time to install: An hour or two.

Where to find it: Heating contractors listed in the Yellow Pages, or your fuel dealer. Technical details of fuel nozzles are available from Furnace and Boiler Test Unit, Brookhaven National Laboratory, Upton, N.Y. 11790. You can also order "How to Improve the Efficiency of Your Oil-Fired Furnace," Technical Information Center, Department of Energy, P.O. Box 62, Oak Ridge, Tenn. 37830.

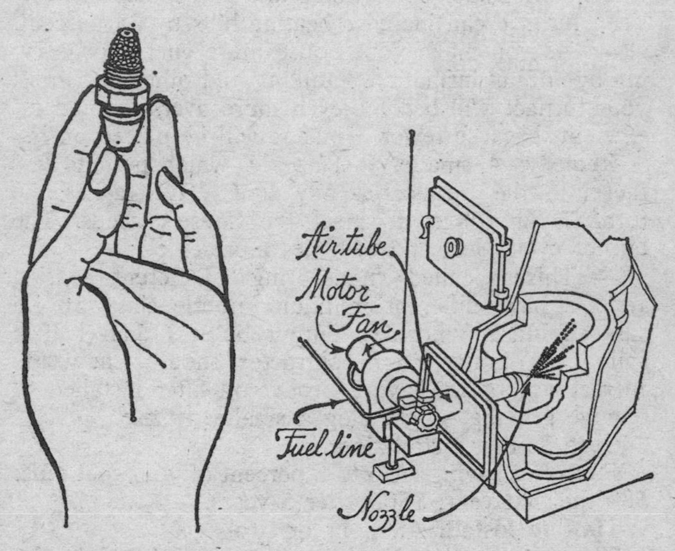

Make your furnace eat less fuel—Give it a smaller mouth.

Tip #12
Reclaiming Furnace Heat

Description: At least 20 percent of the heat you've paid for goes up the chimney. When the furnace is running, combustion gases go up the chimney at temperatures of 400 degrees or more.

— When the furnace is not running, a lot of residual heat goes up the chimney from the still-hot combustion chamber.

Remedies:

— Special pipes or tubes that run through the chimney and connect to your conventional heating system. Called "furnace heat reclaimers," they contain water or air, depending on the type of heating system you have. The escaping heat warms the pipes and the circulating air or water inside. The heat is channeled back into the house.

— A metal plate, called a "flue damper," that automatically closes off the chimney a few minutes after the furnace stops running. The furnace—and ultimately your house—retains heat that would otherwise escape. When the thermostat calls for heat again, the damper opens and the furnace turns on.

Cautions and Advice:

— Buy only reclaimers and dampers approved by Underwriter's Laboratory or some other independent lab. Circuits in approved dampers don't allow the furnace to re-start unless the damper is open. That way, combustion products don't fill the house.

— Have the devices put in by a competent heating contractor. Get estimates. Ask for a list of previous customers and call them. Check the contractor's credentials with the local utility company or trade association of oil or gas dealers.

— The contractor should make sure the escaping stack gases are still hot enough (above 350 degrees) so they don't condense on their way up the chimney. Otherwise, they will corrode the chimney. A thermometer to check should be part of the reclaimer.

— The more efficient your heating system, the less you need these devices. Consider them last, after other remedies such as tuning. Consider a reclaimer if the stack temperature of an efficient furnace is above 500 degrees, a flue damper on oversized furnaces that cycle on and off a lot.

Cost: $150–$250 each.

First-year savings: Where applicable, 5–10 percent of your heating bills, or $100 per year.

Time to install: 2 hours.

Where to buy: Heating contractors in Yellow Pages. Call local office of Underwriter's Laboratories.

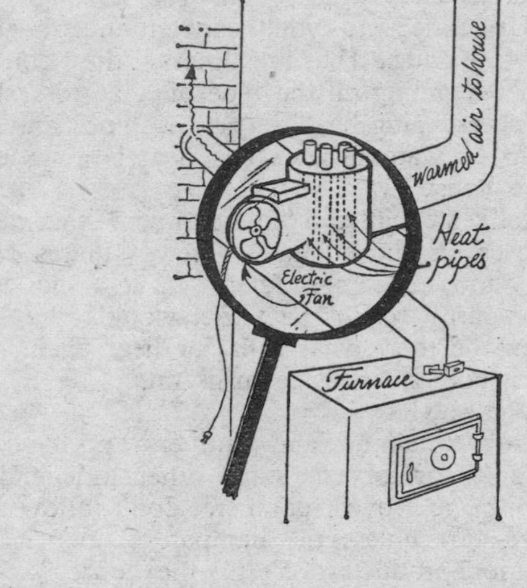

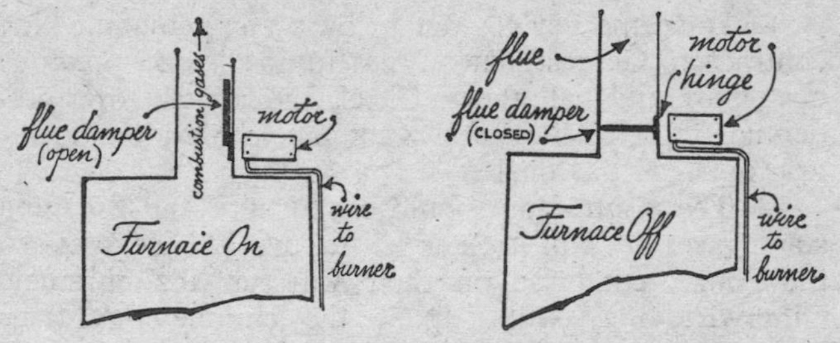

Reclaiming furnace heat.

Tip #13
Three Furnace Fixes

Description: Clogged air filters, fans that run too long and unneeded pilot lights throw your money away for no reason.

(1) *Air filter*—The nose of the furnace. If it is clogged, the furnace won't operate efficiently. It's like trying to breathe with a handkerchief over your nose.

— Check the filter every 30 days. If dirty, replace it.

Cost: $2 to replace.

Savings: Up to $75 a year. Pay-back time—less than a week.

(2) *Fan*—It forces heat up through the ducts of many gas and oil hot air heating systems. The fan is operated by a thermostat in a duct near the furnace.

— Often, the fan turns on at 150 degrees and shuts off at 120, leaving a lot of heat in the furnace. The heat goes up the chimney or into unheated spaces, instead of to your rooms.

— Reset the cutoff for the hot air thermostat (called the plenum or bonnet thermostat) to 5 degrees above room temperature—73–78 degrees. It will run longer and put more heat into your rooms.

Cost: $10 for contractor.

Savings: $20 per year, $125 after five years.

(3) *Pilot Light*—It usually runs all summer on a gas furnace, even though the heating system is off. In winter, it runs continuously, when you only need it to start the burner.

— Turn off the pilot light in the summer and relight in the fall. Or, replace the pilot light with an electronic ignition, which comes on only for a split second to fire the furnace.

Cost: $0 for summer turn-off; $100–$150 for electronic ignition.

First-year savings: Summer turn-off, $15; electronic ignition, $30. Five-year savings: $90 for summer turn-off; $180 for electronic ignition.

Where to buy it: Your fuel company or a heating contractor can perform all the above tasks. Or, a serviceman can show you how to check the air filter and turn off the pilot light yourself.

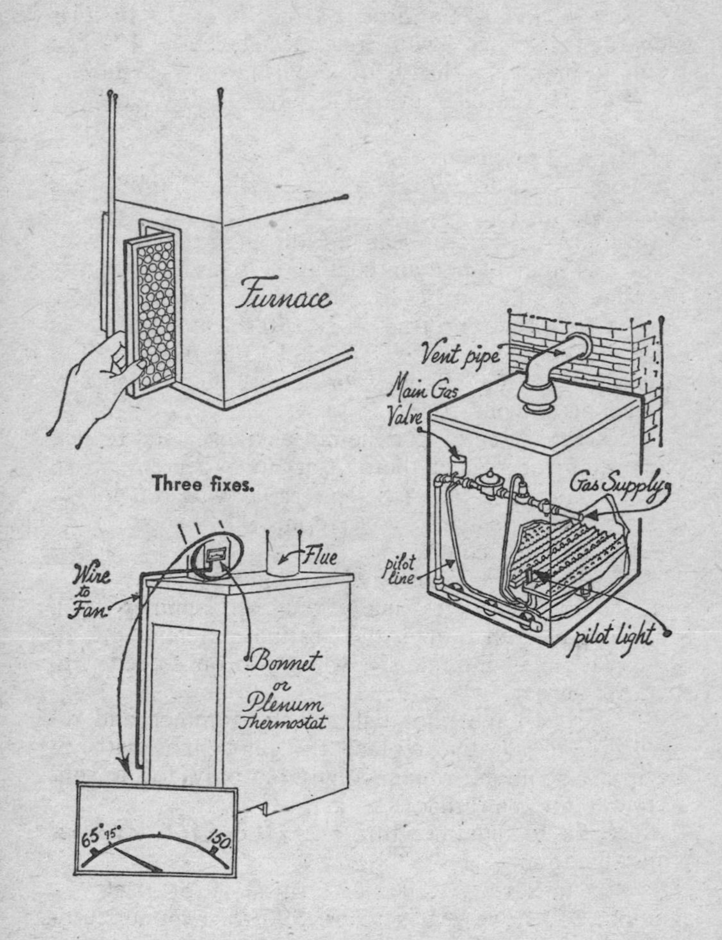

Three fixes.

Tip #14
Overworking the Air Conditioner

Description: Stoves, showers, clothes washers, lights and other heat-producing appliances can increase air conditioning bills drastically. So can improper operation.

— During a 100-day cooling season, appliances can generate the heat in 43 gallons of gasoline and make a central air conditioning system use $100 more electricity.

Remedies:

— Run heat-generating devices early in the morning or late at night. Keep lights, which produce lots of heat, away from the air conditioner. Otherwise, they will create an artifically heated atmosphere near the unit and make it work longer.

— Close drapes to keep out the hot summer sun. Close an air conditioner's fresh-air vent during hot spells. Instead of constantly cooling hot outside air, it will recirculate and filter cooled room air.

— Clean the air conditioner filter once a month. Don't block the air conditioner's circulation. Shade the outside part of the unit by overhangs, bushes or a beach umbrella.

— Don't set the unit's thermostat colder than normal when you first turn it on. It won't cool the room faster. It will only make the room colder than you wish, and waste money.

— The main purpose of an air conditioner is to remove humidity. This is done either at 78 degrees or 72 degrees. Experiment with slightly higher settings.

Cost: $0.

Annual Savings: $20–$30 per room air conditioner, $100–$300 with central system.

Time to implement: Less than an hour.

Air-conditioning—$0 in cost saves perhaps $100 a year.

Tip #15
Cooling the Attic

Description: On hot summer days, the attic can reach temperatures 40 degrees higher than the outside air, or 130–150 degrees. This will put an enormous strain on your air conditioners, as the trapped heat radiates to your cooled living area.

Remedy: A natural draft attic cooling system.

— It is a series of vents at the bottom and top of your roof. Since hot air rises, the heat will rise out of the top vents, letting cooler outside air come through the bottom. This will set up natural convection breezes in your attic.

— The system not only reduces the air conditioning load; it also may reduce the need for more air conditioners.

— In winter, the natural convection removes moisture from the attic, protecting insulation and belongings. With heated attic, close vents partially.

— Even though hot air rises, the intense summer heat trapped in an enclosed attic will get into your living area below by radiating downward—first heating the attic floor and finally the rooms underneath it.

A Word About Attic Fans: No longer are they recommended by energy officials. Numerous studies show that the electricity used by the fans wipes out the savings in air conditioning.

Cost to install natural system: $100–$200.

First-year savings: $20–$30 with two air conditioners; $100 or more with a central system.

Five-year savings: $130 with 2 window units; $600 with a central system.

Where to find it: Heating or air conditioning firm.

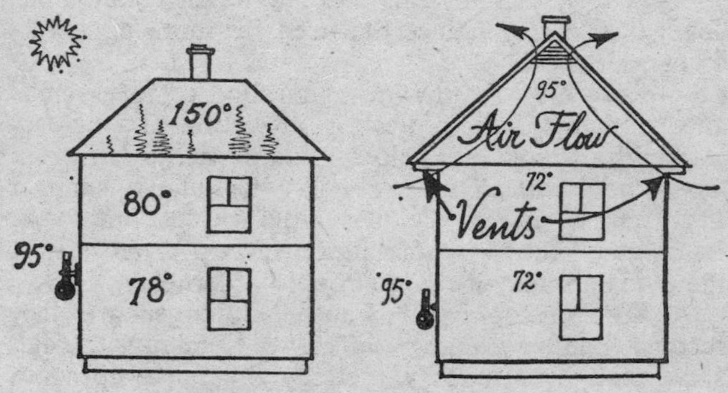

Attic venting cuts cooling bills.

2
Cars

Tip #16
Filling Tires

Description: Put more air in your tires. Save up to $50 per year.

— Low tire pressure may give a soft ride, but it also cuts gas mileage. How? A partially-deflated tire has more rubber in contact with the road, increasing the friction and rolling resistance. It makes the engine work harder to move the car. Like running through water.

— Many car owner's manuals suggest a tire pressure of 22 pounds per square inch. But the maximum inflation—embossed in the rubber on the size of the tire—is 32 pounds per square inch. THAT'S ALMOST 50 PERCENT HIGHER.

— Each 2 pounds of underinflation cuts gas mileage by 1 percent.

Remedy: Look on the side of the tire to find the maximum inflation. Then, when the car is cold (during the first few miles), fill up the tires to a pound or two under the maximum.

— You will get a somewhat harder ride.

Cost: $0.

Annual savings: $50 per year: up to 5 percent of your gasoline costs.

Time to implement: 5 minutes a month.

Where to find it: Gas station.

Epilogue: A Congressional study found that the tires on 4 of every 5 cars are grossly underinflated. If the nation's tires were inflated properly, we'd save 6 million gallons of gasoline *each day,* enough to run all the cars in Philadelphia.

Tip #17
A Car in Tune

Description: You pay for a car tune-up whether or not you have one.

— Without a tune-up, you will pay up to 20 percent extra for gasoline, or $200 a year.

— Bad spark plugs alone can cut gasoline mileage 15 percent. The federal government estimates that one plug misfiring half the time will add 7 percent to your gasoline bills.

Remedy: Tune-ups, according to the specifications for your model car. They will cost $60 and be required every 1½ years for most American cars.

— Tune-ups include changing spark plugs, checking for fuel leaks and blockages, setting the right firing rate of the ignition, adjusting the fuel/air mixture of the carburetor, replacing the air and fuel filters, among other things.

— Power-boats and lawnmowers also need periodic tune-ups.

Cost: $45 per year.

Savings: Up to $200 per year.

Where to find it: Service station.

Epilogue: The average car has 15,000 parts. Things get out of adjustment by normal driving, with bumps in the road and varying weather conditions. An untuned car is also a car with a shorter life.

Tip #18
Idling and Speeding

Description: A car running but not moving and a car moving too fast are everyday, expensive occurrences for which there are cheap remedies.

— An idling car costs 1.5 cents a minute. It uses a cup of gasoline every 6 minutes, a gallon every 96 minutes.

— Driving 1,400 miles at 70 mph costs $32 more in gas than the same distance at 55 mph. At 70, cars use 21 percent more gas than at 55; at 65, 10 percent more gas than at 55.

Remedies:

— When you must idle for more than 30 seconds, shut off the car. Each minute, an idling car uses about 1.4 ounces of gas. But it takes only half an ounce to restart it. The break-even is therefore about 25 seconds.

— Better planning reduces the need for speeding to save time.

— When you first start a car, don't idle for more than 30 seconds. Cars today are meant to be driven almost immediately—even in cold weather—although you should take it easy the first few miles.

Cost: $0.

Savings: Idling 5 minutes less a day: $27 the first year, $150 after 5 years.

— Speeding 100 hours less: $50–$135 the first year.

Epilogue: Idling vehicles are almost an institution: outside of delis while the driver is buying a sandwich; in the middle of streets, while driver and pedestrian converse. Two minutes here, 5 minutes there. A lot of minutes, a lot of money.

Tip #19
Habits

Description: Poor driving habits can double your gasoline bills. Most such habits aren't particularly useful anyway.

— Fast acceleration, constant lane changing and hard braking use up to 50 percent more gasoline. Jack-rabbit starts alone add 15 percent to your gasoline bills.

— Driving at a steady speed uses half as much gasoline as does stop-and-go driving.

— The short trip is the biggest waster of all. The first 5 miles, when the car is cold, use twice as much gasoline per mile as do trips of more than 10 miles.

Remedies:

— Accelerate steadily. Coast to red lights if possible.

— Plan routes to avoid traffic. Saves on your nerves, too.

— Combine shopping, work and other errands to reduce short trips. Even if the errands are in different directions, it will be more economical to do them on one occasion—when the car is warm—than to do them separately, warming up the car each time.

— Use the Yellow Pages. Phone ahead for unusual items.

Cost: $0.

Annual savings: As much as $500.

Epilogue: Some figures, based on various studies—(1) Weaving in and out of traffic over a single 50-mile trip saves about 5 minutes and costs $4.60 extra in gas. (2) Driving 1,000 miles in individual 1-mile trips uses 4 times as much gasoline as the same 1,000 miles covered in 40-mile trips. Extra gasoline cost for the 1-mile trips: $165.

Gains a few minutes a day, costs $150 a year.

Tip #20
Octane and Shopping for Gas

Description: Using higher octane gas than you need does nothing for your car and depletes your pocketbook. Shopping around for lower-priced gas saves far more than you might think.

— A gallon of gasoline has the same amount of energy no matter what grade it is. Higher octane gasoline is only for higher-compression engines. It does not give more power to cars designed to use lower-octane gas. It merely costs their owners 10 percent more for gas each year.

— Gas prices vary as much as 20 percent from station to station in many localities.

Remedies:

— Match the octane on the pump to the octane your car needs, found in the owner's manual or by asking a dealer.

— Patronize self-service stations or those with lower gas prices. Find out when they're open; phone ahead to find out if there are lines. A little extra time and organization will save you money and make you feel more in control.

Cost: $0.

Savings: $100 per year if you are now using the wrong octane; $200 per year if you are buying the most expensive gas around.

Epilogue: About 5 percent of the cars need premium gasoline, yet 25 percent of the gas sold is premium. This costs American motorists more than $1 billion each year.

Tip #21
Oil and Gas

Description: Oil is the blood of the engine. It cools, lubricates and removes impurities from the air and gasoline. Over time, it picks up sand and grit.

— If not changed regularly, it will reduce gas mileage and eventually rub off parts of the engine, shortening car life.

Remedies:

(1) Change the oil and oil filter as recommended in the owner's manual, usually every 5,000–6,000 miles. If you do a lot of stop-and-go driving, however, change the oil and filter every 2,000–3,000 miles.

Cost: $100 per year.

Savings: $50 a year in gasoline, plus tens of thousands of miles on your car.

(2) Buy oil in bulk at a home improvement center if your service station will change it under those conditions.

Cost: $0.

Savings: $30–$35 per year.

(3) If you have a new car, consider long-life, synthetic oils. They last 25,000 miles or 1 year, whichever comes first. (Change the oil filter every 6,000 miles in any event.) Not for older cars that burn or leak oil, unless you like losing $4-a-quart fluid. Make sure a synthetic oil won't affect your new-car warranty.

Cost: $30 for a year of normal driving (12,000 miles).
$36 for 25,000 miles in a year.

Savings: $50 a year in gas (5 percent) for normal driving;
$100 a year for twice-normal driving, plus $25 in oil.

The more you drive, the more you save with long-life oils.

Tips #22
Air Filters, Alignments and Gadgets

(1) Description: A dirty air filter can add 10 percent to gasoline bills.

Remedy: Check the air filter once a month. It is the large round object on top of your engine. Unscrew the wing-nut and bang out the dirt from the filter, or replace it.

Cost: $2.

Annual Savings: $100 if continuously dirty now.

(2) Description: Unaligned front wheels and brakes can add 3 percent to your gasoline bills.

Remedy: Align brakes and tires.

Cost: $50 when needed, perhaps every 1½ years.

Savings: $30 a year, plus your car or your life.

(3) Gadgets and Additives: Two words are appropriate—be careful. Many salesmen claim 25 percent gasoline savings. Tests on several gadgets by Consumer Reports found no significant change in gasoline mileage. To test claims that sound good, ask for test results from a reputable, independent laboratory. Consider gadgets only after you've done everything else.

Epilogue: Experts agree that (1) most of the 40 million yearly car breakdowns are due to lack of normal maintenance, and (2) 90 percent of all car repairs could be avoided by periodic check-ups. Careful maintenance will give you a 150,000-mile car, saving thousands of dollars in new-car bills and perhaps thousands more in break-downs and aggravation.

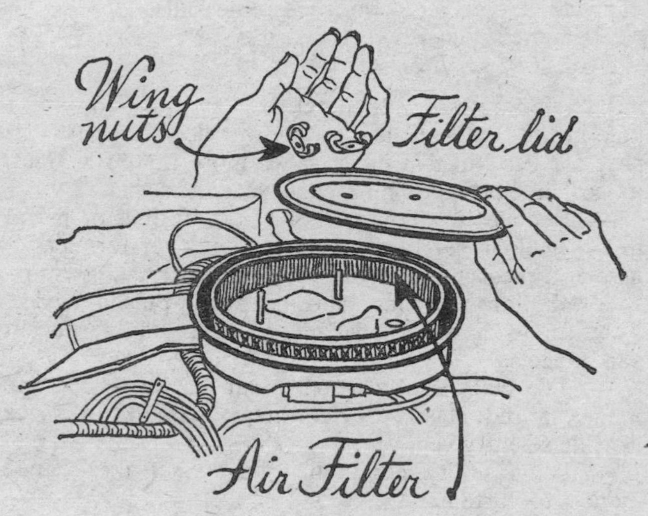

Costs $2, saves up to $100.

3
Hot Water

Tip #23
Pinching Hot Water

Description: A small metal or plastic gizmo with a hole in the center. Called a "shower flow restrictor," it pays for itself in two days; saves more than 100 times its cost in hot water bills each year.

— Reduces the flow of hot water by half or more: to 2–4 gallons per minute. Plenty left over for cleanliness and pleasure.

Installation: Unscrew shower head, put in device. Then screw shower head back on. Can also be put on sink faucets.

— The U.S. Energy Department has sponsored one model, a red plastic device that looks like a space capsule re-entry vehicle.

— Some utility companies have sent disc-shaped models to customers.

— Try each one out. They come with instructions.

Where to find it: You will probably have difficulty finding it at local hardware stores, although the companies that make them are trying to expand distribution. Most stores sell only $2 valves or complete replacement shower heads for $8 or more. You can get the devices by sending money and a self-addressed, stamped envelope to the following places:

— W.F. Products, 3625 Dividend Drive, Garland, Tex. 75042. (red plastic device, 35 cents.)

— Conservation Awareness, 121 West Oak St., Amityville, N.Y. 11701. (metal disc, 25 cents.)

Cost: $1.50 for a whole house.

Savings: Up to half your hot water bills, or $180 a year.

Time to Install: 5 minutes per outlet.

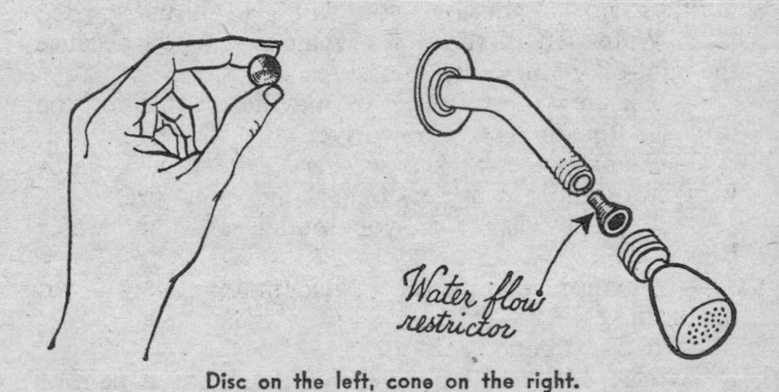

Disc on the left, cone on the right.
Saves a hundred times its cost each year.

Tip #24
Nipping the Drip

Description: A drip from the bathroom faucet can fill a cup in 10 minutes and waste 3,260 gallons of hot water per year.

— To heat that water it took $10 worth of natural gas, $22 in oil or $40 in electricity. All down the drain.

— Many leaky faucets in tubs use 2–3 times that much water; many houses have more than one leaky faucet.

Remedy: A small rubber washer and a few minutes' time.

— Turn off the water by using a shut-off valve underneath the sink or the main inlet in the cellar or utility room. Otherwise, you will get an unexpected bath. Water left in the pipes won't squirt you because the shut-off reduces the pressure.

— Unscrew the faucet. You may have to pry off the cap with the tip of a screwdriver.

— Remove the broken washer. Take it to a hardware store and buy a new one of the same size.

— Replace washer on your outlet. Turn the water back on.

— No more drippy faucet. No more money down the drain.

Cost: 5–10 cents per outlet.

Savings: $10–$100 per year. Enough for a pair of sneakers, or maybe even a weekend vacation.

Time to install: 10 minutes or less per outlet.

Epilogue: The U.S. has about 300 million residential faucets and several hundred million in hotels and other commercial buildings. About 10 percent of all faucets leak. Back-of-the-envelope calculations suggest that 150 billion gallons of hot water are dripping needlessly each year, wasting about $1 billion in fuel.

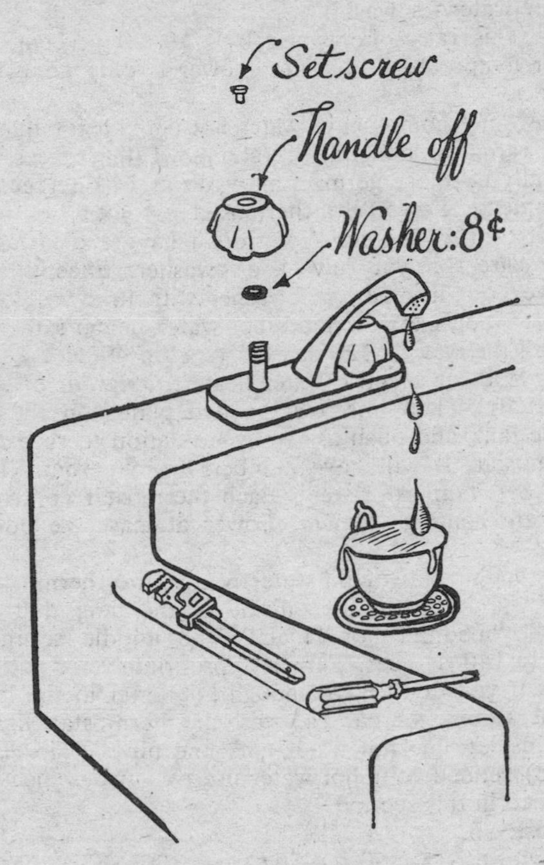

Why drip money down the drain?

Tip #25
Too Hot Water

Description: Your water heater may be warming water to 180 degrees, when you need only 120–140 degree water for washing.

— This raises hot water bills 10–20 percent. The water temperature of a hot shower is only about 105 degrees.

— Many homes have water hot enough for quarantine wards in hospitals. Water at 170 degrees kills virtually no more germs than water at 140 degrees.

Remedy: Lower the thermostat on your hot water heater to 120 degrees if you don't have a dishwasher, 140 degrees if you have a dishwasher. Exception: If you've bought a new dishwasher with its own electric booster, you can lower the hot water heater's thermostat all the way to 120 degrees (see tip 49).

— With an electric water heater, *first shut off your electricity*. Then unscrew the two plates on the side of the tank and push aside any insulation to reveal the thermostat. It will have numbers from perhaps 110–180. Set it appropriately. Each thermostat controls a separate heating element. Lower at least the bottom one.

— With a gas or oil water heater, the thermostat is usually a knob on the outside of the tank that says "low," "medium" or "high." The middle setting is usually 140 degrees. Turn it appropriately.

— If you don't have enough hot water at the lower temperatures, you can (a) raise the thermostat slightly, (b) insulate the hot water tank and pipes (Tip #26) or (c) reduce your hot water use by various measures outlined in this section.

Cost: $0.

Savings: $15–$60 the first year; $90–$350 after five years.

Time to implement: Five minutes or so.

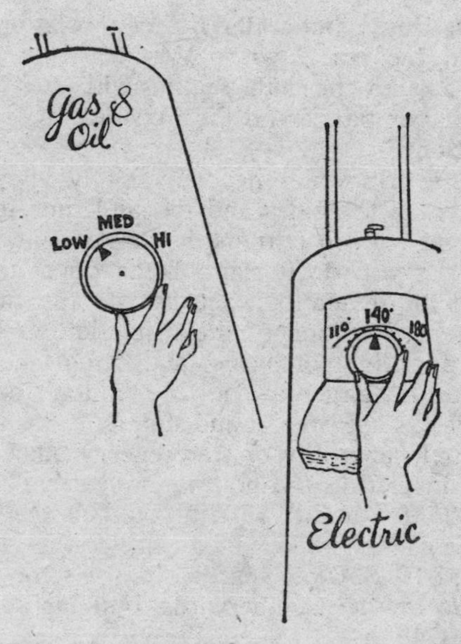

Your home is not a quarantine ward. Lower hot water temperature and save money.

Tip #26
Bundling Up the Hot Water Tank

Description: Just as a blanket or coat keeps in body heat, insulation wrapped around a hot water tank keeps the water inside warmer, reducing fuel bills up to 10 percent.

— Touch the surface of your present hot water tank. If it's warm or hot, you need insulation. Many models have insulation inside, but most don't have enough.

Remedy: Buy a hot water heater insulation kit. Or, for a third the price, wrap 3½-inch-thick fiberglass insulation around the tank, sealing with duct tape. Put the silver vapor barrier on the outside.

Installation:

— On electric water heaters, simply wrap the insulation around the sides and top and tape it.

— On gas and oil burners, it's trickier, and a safety hazard unless you do it right. Don't cover the top of the tank with insulation, because of the air intake. Also, cut out a section of the insulation so it doesn't touch the pilot light or burner and catch fire. Remember: Driving is dangerous too, if you don't do it right.

— Refit kits are pre-cut and simpler to install.

— Your local utility or state energy office has brochures on insulating the hot water heater.

Cost: $25 for a kit; $10 if you buy the materials yourself.

Savings: $10–$30 per year.

Where to buy it: Lumberyards, building supply outlets, Sears. Call first.

Time to install: 1–2 hours.

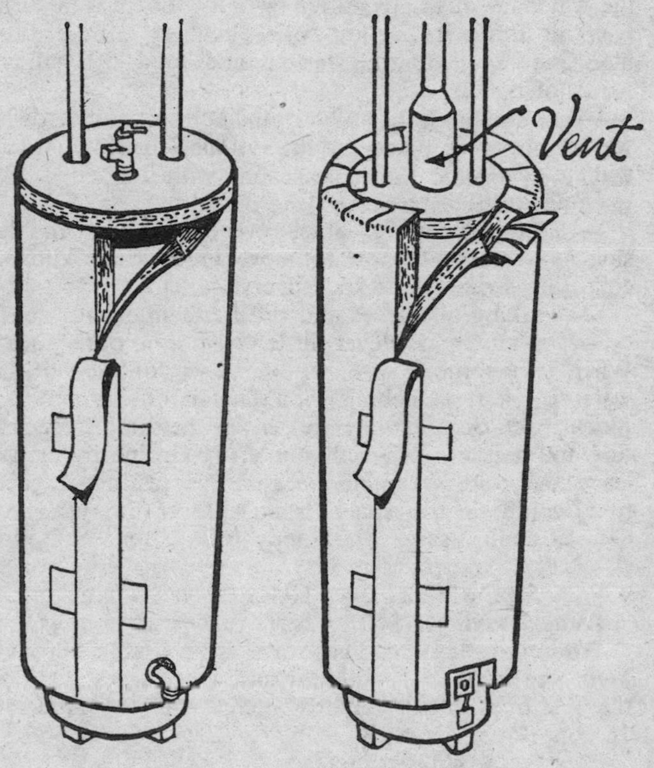

Wrapping the hot water heater.

Tip #27
Washing Dishes Cheaper

Description: Whether by hand or machine, most people do not wash dishes efficiently.

— Time-and-motion studies show that people washing dishes at the kitchen sink let up to 70 percent of the water go down the drain without touching a dish. Running the warm water costs about ¾ cent a minute. Over a year, the waste could total 50 gallons worth of heating oil.

— The average family with an automatic dishwasher does 416 loads a year, costing $80 in hot water and electricity.

Hand-washing remedies:

— Better planning: Stack everything next to the sink so you don't have to move around the kitchen collecting stray plates and silverware.

— Soaking all dishes and rinsing them off at once.

— Attaching a clever little valve to your faucet. When you scrub a pot or otherwise don't need the water for a few seconds, you flick it off. When you flick it back on, the water returns at the same temperature and pressure at which you left it. Ordinarily, you'd leave the water running because readjusting it each time would be too much trouble. Over the year the device could save thousands of gallons of hot water.

Cost: $0 for habits, $4–$10 for the valve.

Annual savings: $40 or more in hot water bills.

Where to find it: The valve is available through Montgomery Ward and local hardware stores. One of the best ones is the "Hydroswitch," made by Abacon, 16 Pont Street, Great Neck, N.Y. 11021.

Machine-washing remedies:

— Wipe soil off dishes with the paper napkins you used at dinner. Washing the dishes off at the sink first defeats the purpose of a dishwasher.

— Wash only full loads. You use the same amount of water whether you put in one dish or 30.

— If the dishes are only lightly soiled, use a short cycle.

— Consider skipping the dry cycle and save the

energy used by the heating element. Let the dishes air dry.

Cost: $0.

Savings: $20 a year by using a short cycle.
$20 a year by doing only full loads.
$6–$8 a year if you air-dry the dishes.

Epilogue: Used properly, automatic dishwashers consume less water than normal hand-washing of dishes.

Tip #28
Washing Clothes Cheaper

Description: A washing machine uses up to $70 in hot water each year. About half of that is unnecessary.

— A hot wash with a warm rinse uses 24 gallons of hot water.

— A warm rinse is never necessary. It's no better than a cold rinse for washing off soap. Virtually no one who understands washing machines recommends a warm rinse.

— Except for very soiled clothes, a hot wash is unnecessary: A warm wash will do the job.

— Residential washing machines never sterilize clothes: You don't kill more germs with a hot wash than a warm wash.

Remedy: Wash clothes in warm water except when highly soiled. Always rinse clothes in cold water.

— If you use a cold-water detergent and your clothes are not very dirty, they might clean quite well with a cold wash and a cold rinse.

— Experiment. Use the coldest water you can to get the job done. Why throw away money to heat water that doesn't get your clothes any cleaner?

Cost: $0.

Savings: 7–10 cents a load by switching from warm to cold rinse: $20 the first year; $125 after five years.

— Twice these savings by switching from hot wash/warm rinse to warm wash/cold rinse. Three times the savings by washing completely in cold water.

Epilogue: If two-thirds of the residential washes were in cold water, the U.S. would save enough energy to heat a million homes a year.

4
Cooking

Tip #29
Snuffing Out the Pilot Light

Description: The pilot lights on your gas range consume half the energy (and money) used for all your cooking!

Remedy: Get as many as possible disconnected by an appliance repairman, who will turn the appropriate valve. DON'T DO IT YOURSELF.

— On most stoves, the pilots on range-top burners can easily be shut off. Then you can use matches, a hand-operated sparking device or a battery-powered device with a long nozzle: You press a button and the nozzle's tip lights the stove.

— Disconnecting the oven pilot light presents problems, because the oven cycles on and off during operation. Without the pilot, there would be nothing to ignite the gas when the oven cycles on again. Older stoves built without pilot lights had different cycling mechanisms, which reduced the flame but didn't turn it off. Reworking a newer model to this method is quite expensive and repairmen are reluctant to do it.

— The range-top pilots use 2–3 times as much gas as the oven pilot anyway.

Who to call: Look in the Yellow Pages under "Appliance Repair-Major Appliances." Sparking devices are sold at hardware or appliance stores, also found in the Yellow Pages.

Cost: \$14–\$20 for the repairman to disconnect the pilot light. \$2–\$8 for a sparking device.

Savings: $10–$20 the first year, $60–$125 after five years.

Epilogue: Disconnecting range-top pilot lights alone would save enough energy to heat 3 million homes a year. Some may consider this a sacrifice, but it is included here because many people ask about it.

A battery-operated sparking device lighting a stove.

Tip #30
Cooking on the Range

Description: Boiling a large pot of pasta once a week without a lid on it could cost $8 a year extra in fuel bills. Boiling water for six cups of tea each day when you drink only two cups could cost as much as $14 a year extra.

— Usually, only 40–50 percent of the energy in the range-top burners goes into the food you eat. You can increase this percentage, and save money, by better cooking habits.

Remedies:

— Cover pots when boiling water. The lid will keep the heat in and cooking time will be about 20 percent shorter.

— Don't boil more water than necessary. Why pay for the extra energy for nothing? Try steaming vegetables: It uses less water and is more healthful. If you have a pressure cooker, consider using it more often: Cooking times are even shorter.

— Size the pot for the burner. A small pot on a large burner allows wasted heat to escape around the sides.

— Clean the bottoms of pots well so the heat transfer into the food is better. Clean the heat reflector under each burner so heat bounces back onto the pots.

— On a gas range, check the flame. A blue flame means the burner is working properly. A yellow flame means food particles may be clogging the jets.

Cost: $0:

Savings: $5–$20 per year; $30–$125 after five years.

There are at least 3 things wrong with this practice.

Tip #31
Cooking in the Oven

Description: One family can use twice as much energy (and money) as another family to cook the same foods on the same oven. Why? Poor cooking habits.

Remedies:

— Don't peek into the oven unnecessarily while a meal is cooking. Each time you do, 20 percent of the heat escapes. Doing it five times per meal doubles the cooking cost. Peer in through the glass oven door if you have one.

— Try to cook entire meals at once. Although you must open the door to remove foods with different cooking times, you still save $4–$9 a year. Each time you cook an entire meal at once you spend 6 cents. But cooking each dish separately, either in oven or on range-top burners, costs 9–12 cents.

— Turn off the oven 10–15 minutes ahead of time and let the residual heat finish cooking the meal. Saves half an hour of cooking time per week, about $2.25 a year.

— Ceramic and glass ovenware allows oven settings 25 degrees cooler than metal pans, because they retain heat better.

— Divide large breads, meat loafs and other foods in halves or quarters, cutting cooking time by exposing more surface area to the heat.

— Since the oven warms an entire cavity, it doesn't make sense to heat all that space to cook two potatoes. If you have a toaster oven or microwave, use it for smaller items.

— If cooking instructions allow it, thaw frozen foods before putting them into the oven. It will save the fuel necessary to defrost the foods inside the oven.

Cost: $0.

Savings: $10–$30 per year.

Epilogue: Throughout America, children peek into ovens 7 times in 12 minutes to check the chocolate chip cookies.

Entire meal cooked at once, saved 6 cents. It adds up.

Tip #32
Food Coolers

Description: The average refrigerator door is opened 25 times a day. But one study found a family's food cooler being opened 100 times daily. Why? A child was taking grapes—individually—from the refrigerator.

— Extra door openings can add 15 percent—$15—to your yearly refrigerator bill.

— A refrigerator costs $75–$100 a year to run. But you can double or halve that cost, depending on your habits.

Remedies:

— Don't stand with the refrigerator door open for 2 minutes, thinking, as the unit cools the room.

— Check the door seal (gasket) by opening the door, putting a dollar bill between the seal and the frame, and closing the door. If it's hard to pull out the bill, the seal is tight enough. If it's casy, the seal is too loose. Re-glue it or buy a replacement at an appliance parts store. Check the gasket in a few places.

— Disconnect the old, rarely-used second refrigerator in basement or garage and save up to $100 per year.

— Try to keep the refrigerator away from heat sources such as a stove, direct sunlight or ducts. The heat will make the unit work harder to cool the foods inside.

— Cool hot foods at room temperature before refrigerating. Clean the condenser coils at the back or bottom of the unit at least twice a year.

— Don't set thermostats too low. For a refrigerator, 32–40 degrees is fine; attached freezer, 10–25; separate freezer, 0–5.

— Set power-miser switch on "dry;" energy-saver switch to "on." It will turn off heaters that evaporate moisture from the side of the refrigerator. They are unneeded except on very humid days in houses without air conditioning. Savings: $10–$16 per year.

Cost for all measures: $0 ($5 for a new gasket).

Savings: $25–$100 or more per year, depending on how wasteful you are now.

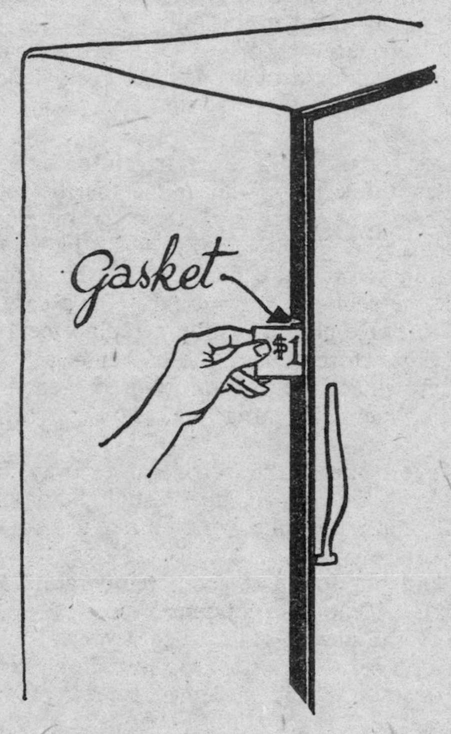

If it pulls out easily, dollars are leaving your refrigerator with the escaping cold air.

5
Lighting

Tip #33
Gas Lanterns

Description: Ornamental gas lanterns are among the world's most wasteful energy users. They use more than 20 times as much energy as their replacements, 25-watt electric bulbs.

— Many lanterns burn all the time, costing $100 per year.

— Various states prohibit ornamental gas lanterns, and the devices will be illegal throughout the country in 1982. But officials concede that policing will be difficult: The entire nation would have to be inspected. Consumers will have to look out for their own economic welfare.

Remedy: Have a plumber disconnect the gas lantern. It's usually just a matter of closing a valve.

— An electrician can run a wire to the lantern and install a wall switch inside your house. It won't be cheap, but it will be cheaper than letting the gas burn.

Cost: About $20 to disconnect the gas, $40–$100 to install an electric bulb with a wall switch.

Savings: $40–$100 the first year; $250 to $600 after 5 years.

Epilogue: Estimates of the number of ornamental gas lanterns in the U.S. range from 1.5 million to 6 million. The wasted energy could heat at least several hundred thousand homes. That's a cheaper alternative to expensive natural gas imports from Canada, Mexico and other areas.

Tip #34
Good Bulbs, Bad Bulbs

Description: All incandescent lightbulbs are not equal. Choosing and using the wrong bulbs will cost you needless money.

— Many lighting fixtures accommodate too many bulbs, meaning you pay to light 4 bulbs when you often need only 2.

— Higher-watt bulbs are far more efficient than lower-watt bulbs. A 75-watt bulb gives off 68 percent more light than three 25-watt bulbs. A 100-watt bulb gives off 20 percent more light than two 60-watt bulbs.

— Long-life bulbs emit 20 percent less light than their standard counterparts.

Remedies:

— Replace some of the bulbs in your over-bulbed fixtures with burned out ones for safety.

— Use high-watt bulbs whenever possible (but not more than the fixture allows).

— Don't buy long-life bulbs, except for hard-to-reach places and in areas where you don't mind the light reduction. It takes five 100-watt long-life bulbs to equal the light from four standard 100-watt bulbs.

Cost: $0.

Savings: About 25 percent of your lighting bill, or $18–$25 per year.

— In a single four-bulb fixture, using only two of the bulbs can save $3 per year; $20 after 5 years.

— All things considered, long-life bulbs cost 15 percent more than standard bulbs.

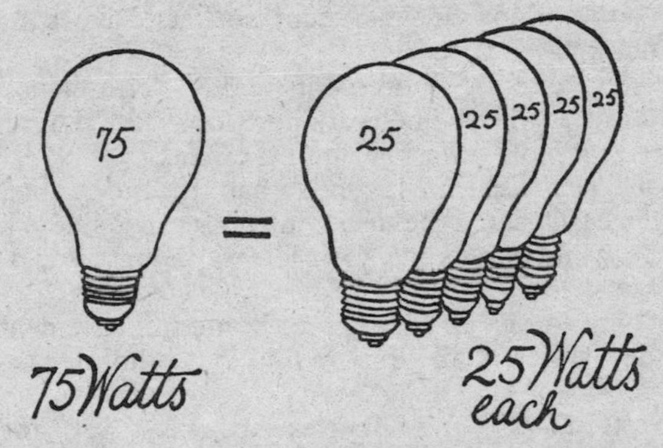

Higher-watt bulbs are more efficient than lower-watt bulbs.

Tip #35
More Efficient Lights

Description: A standard 100-watt lightbulb burning for 10 hours uses as much energy as a pound of coal. Only 5 percent of that energy is converted to light. The rest is released as heat.

— It will cost $20 per year to operate such an inefficient bulb. And you will have to buy four of them, because each will burn out in 3 months.

Remedy: Fluorescent bulbs, which use a third the energy as incandescents to produce the same amount of light. Fluorescents also last up to 10 times longer.

— Recently, circular fluorescent tubes that screw into conventional incandescent sockets have come on the market.

— The circular fluorescents, called "circ-lights," are 8–10 inches in diameter. They look like fluorescent tubes that have been bent into a circle.

— Their light is harsher than incandescents, but they clearly could be used in work and study areas, kitchens and basements. (For lamps, though, you'd need a large shade.)

Where to buy it: Hardware, home improvement and lighting stores. Call first to find out who stocks circ-lights.

Cost: $7–$15 per fluorescent, plus $20 for the electricity to run it for 5 years. Total $27–$35.

Savings: $35–$42 per fixture. It would cost $70 for 10 incandescent bulbs plus the electricity to run them for 5 years.

Epilogue: These costs and savings are based on average usage. If we used the high figures mentioned at the beginning of this tip, a single fluorescent would be at least $70 cheaper than a succession of 20 incandescents over a five year period.

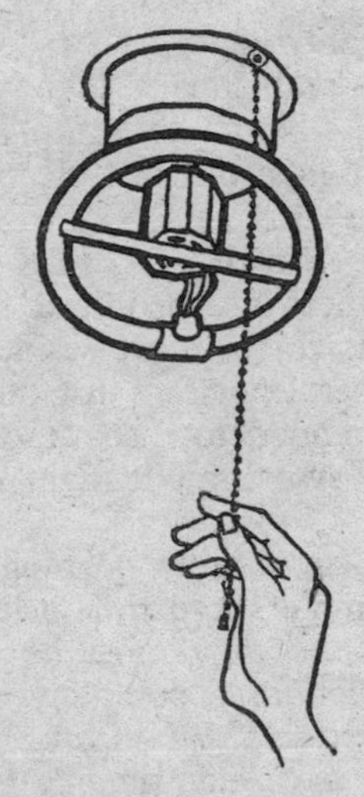

New fluorescents fit into standard sockets.

Tip #36
Dimmers and Timers

Description: (1) You go to work in the morning and don't want to come back to a dark porch. So you leave the outside lights on all day. (2) You have company and don't want most of the house to be dark. So you light empty rooms.

Remedy: Dimmers and timers.

— Dimmers allow a whole spectrum of wattages, saving perhaps half the cost of operating each lighting fixture. Often, when you turn on lights in a room, you neither need nor want the full wattage. Dimmers also allow for romantic ambience. Make sure you buy a "solid state" model, which cuts lighting and electricity. The others reduce lights, but not electricity use.

— WARNING: DO NOT USE STANDARD DIMMERS FOR FLUORESCENTS. IT'S A FIRE HAZARD. Special dimmers are required for fluorescents. They are too expensive for most residential use.

— Timers are useful not only for outside lights, but for inside lamps intended to discourage burglars. The lamps automatically switch on at night when you're not home.

— Lamp timers can simply be plugged into a wall socket. But both dimmers and timers for outside lights connect to switches, so they must be wired in place. Do-it-yourself instructions accompany each device—or you can hire an electrician. When hiring, try to arrange all work at once to save electrician's fees.

Where to buy it: Electrical supply outlets, home improvement centers, hardware stores.

Cost: $5–$10 each for dimmers; $10–$20 each for timers.
$20–$50 for an electrician.

Savings:

— For 3 dimmers—$10–$15 a year; $60–$90 after 5 years.

— Timers on two lamps—$7.50 a year; $45 after 5 years.

— Timer on one outside bulb—$9 a year; $50 after 5 years.

— Timer on outside spotlights—$30 a year; $180 after 5 years.

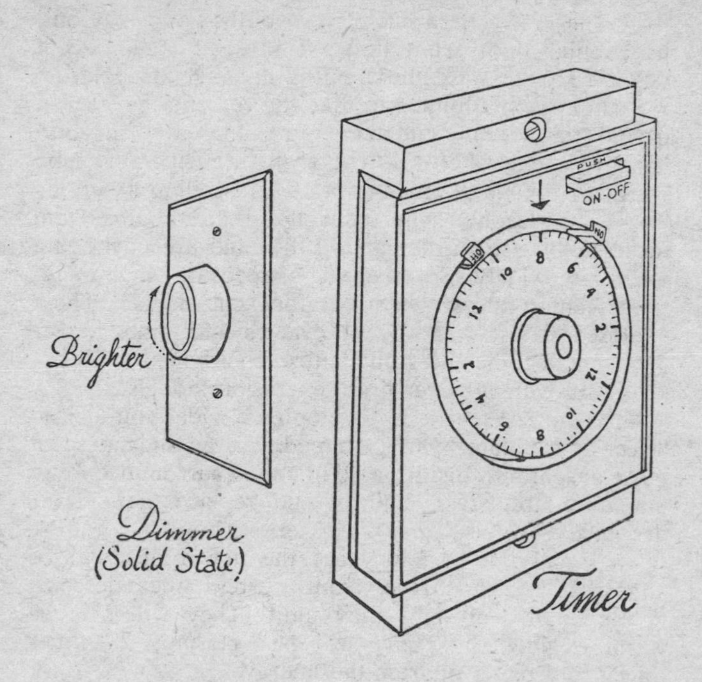

Tip #37
Future Lightbulbs

Description: A major research effort is underway to develop lightbulbs far more efficient than any being used by consumers today. The goal is to reduce individual lighting bills by half—or more—in the next few years.

Specifics:

— The first generation of new bulbs is already out, the circular fluorescent that fits into an incandescent socket (Tip #35). These bulbs are an extrapolation of existing technology. Circular fluorescents have been around for years, but they were unable to fit into standard sockets before. One problem is that the tube is too big for many existing lamp and ceiling fixtures.

— The second generation fluorescent bulbs were successfully test marketed in 1981, and are now being sold more widely. Some are U-shaped and fit into incandescent fixtures too small for "circ-lights." These fluorescents last up to 12,000 hours—15 times longer than standard incandescent bulbs. They reflect a 50 per cent improvement over first generation models.

— General Electric is developing a widely-publicized "electronic halarc" bulb, shaped like an incandescent but transparent. Inside, a filament is surrounded by a miniature arc. Each bulb would replace 5–7 incandescents.

— Finally, in a few years the new bulbs will be almost indistinguishable from current incandescents in the quality of light they emit. They will give as warm a glow. But they will last perhaps 20 times longer and use a quarter the energy.

Cost: $7–$10 per bulb—10 times the cost of current bulbs.

Savings: $35–$50 per bulb over its life.

Where to buy it: Pay attention to news articles on lighting advances. Ask about the advanced bulbs when you happen by a lighting store. There are now fluorescent tubes on the market that draw 20 watts of electricity but give off as much light as a 75-watt bulb.

Epilogue: Get the satisfaction of being among the first to understand and use new technology.

6
Woodburning

Tip #38
Not Heating the Sky

Description: A fireplace flue can draw enormous amounts of extra heat from your house, whether or not a fire is burning. That's because room air—preheated by your conventional heating system—is drawn up the chimney and into the sky.

— With a large fire, all the air in a 10-by-12 foot room goes up the chimney every 2 minutes. This warm air is replaced by cold outside air forcing its way through small cracks in window frames, doors and other areas. Your heating system warms the new cold air, which in turn goes up the chimney.

— Notice how other rooms in the house cool down when a fire is burning. In fact, fireplaces often waste more energy than they give, meaning they *deplete* 10–15 percent of your home's heat. At best they add 10 percent of the wood's original heat. The rest goes up the chimney.

— Even when a fire is not burning, an open chimney flue can let out 8 percent of your home's heat on a continual basis.

Remedies: When a fire is burning, lower your thermostats to 50 or 55 degrees to reduce the operation of your heating system. Close doors between the rest of your house and the room with the fire, to reduce the flow of warm air up the chimney.

— Buy a damper—a hinged metal plate—for your chimney flue. When not using the fireplace, close the damper tightly—and check it. Many supposedly closed dampers have inch-wide gaps for heat to escape. If your current damper is loose when closed, seal with asbestos tape.

— A notched damper will let you open it wide when you start a fire, and then cut back as the fire gets going. Less heat will be directed up the chimney, more into your house.

— Try to burn out the fire before going to bed: A damper open all night can double a house's heating needs. If you can't burn out the fire, leave the damper open only a small amount.

Cost: $0 for habits, $100–$200 for a high-quality damper, installed.

Savings: Sealing your current loose damper, $15–$50 a year. Installing a damper, $100 a year. Operating fire correctly, $100–$150 a year (includes extra wood you save).

Where to buy it: Stores that sell wood stoves; lumberyard. Measure your chimney first. You can have a damper installed by a specialist found in the Yellow Pages under "chimneys."

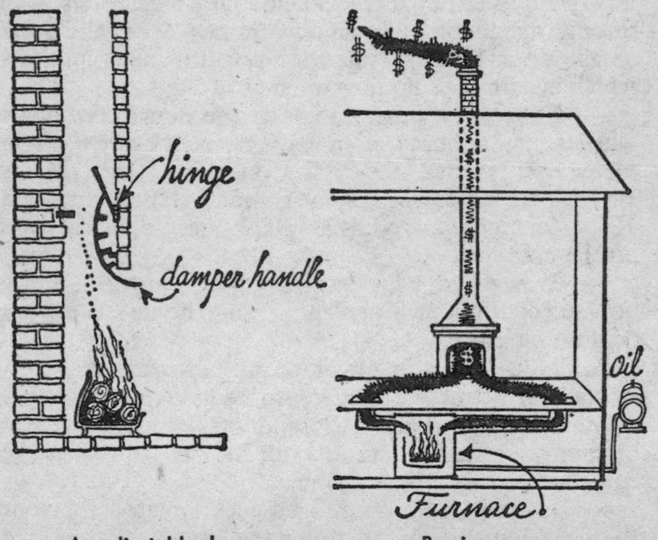

An adjustable damper. Burning money.

Tip #39
Buying Wood, Burning Wood

Description: Buying and burning wood incorrectly wastes hundreds of dollars. Each year, many people get less wood than they pay for. They complain to officials, but usually have insufficient proof.

Remedies: Understand the units. A cord of wood is 128 cubic feet—a stack 4x4x8 feet. Always have it stacked, and measure it. Get a receipt listing the units, amount you've paid and the merchant's identity.

— In general, buy hardwood—oak, maple, hickory—stored at least a year to become seasoned. Unseasoned, green wood smokes and produces 15 percent less heat, because the wood has a lot of moisture. Exception: If you can store it, consider green wood. It's cheaper.

— Comparison shop. Wood prices may vary 20 percent in your locality. Try to buy in the off-season, saving another 20 percent. Unless you're extravagant, don't buy single logs in supermarkets. They cost 3–4 times as much as cord wood.

— Buy several cords at once if you can, either alone or with neighbors. The merchant may give you a discount. In any event, your supply will become inflation-proof.

— If you have a means of transporting it, buy wood as close to the forest as possible. It may halve your wood bill.

— Scrounged wood is cheapest but takes the most work. It includes tree branches pruned by your town, and scrap from lumberyards and landscapers. The U.S. Department of Agriculture office can direct you to state and national forests that allow free wood-cutting by the public.

— It's not clear that grates of curved hollow tubes justify their extra cost in extra heat. In theory, they draw cool air in and blow hot air out. In practice, many wear out quickly. Get a 5-year guarantee against heat wear or don't buy.

Where to buy: Look in newspaper classified ads and shoppers' throwaways. Ask at wood stove stores for the best buys. In rural areas also ask at hardware and general stores.

Cost: $0 in most cases. Axes cost $25; chainsaws, $200.

Savings: Wood averages $100–$125 a cord. Comparison shopping can save $25 a cord; bulk buying, $40–$50 a cord. Burning five cords a year, you can cut your wood bill from $600 to $300.

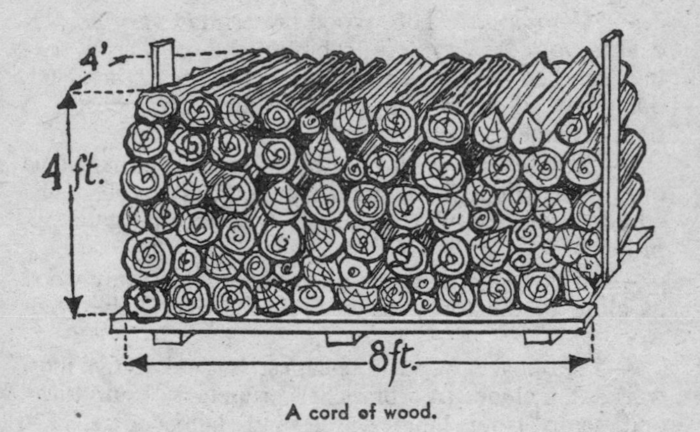

A cord of wood.

Tip #40
Paper Logs and Coal

Description: Coal and tightly-rolled newspapers can extend your wood supply. Newspapers can replace half the wood you would otherwise burn. Alternating between wood and coal lets you take advantage of market conditions.

Newspaper logs: Roll the papers tightly by hand and bind with wire. Or buy a paper log roller: A metal trough with a rod inside and a crank at one end. You place a paper around the rod and turn the crank.

— Newspaper logs have three-fourths the heating value of wood. You can't make a complete fire out of the paper logs: They will just smolder, because various burnable components have been processed out. First get a strong wood fire going, then use two paper logs for every wood log. Works in both fireplaces and stoves.

Cost: $0 by hand; $15–$25 for a paper-log roller, sold in stove stores and some home improvement centers. Call first.

Savings: $25 for a year's worth of one daily newspaper, which contains the heat of a quarter cord of wood.

Coal: Burns with a much hotter flame than wood. NEVER use it in a stove not rated for coal: It may cause a fire. For a fireplace, get a heavy, basket-shaped iron grate.

— Start a coal fire with a small wood fire. Anthracite—hard coal—is best for stoves. In fireplaces, use anthracite or easily-ignited blocks of cannel coal that emit flickering, luminous flames. Cannel comes from the Welsh "canwyl," meaning candle.

— Per unit of heat, coal is half the cost of oil and electricity. It's double the cost of natural gas. To compare with wood, add 25 percent to the price of a cord. If a ton of coal costs less than that figure, coal is cheaper.

— Look in the Yellow Pages under "Coal." The merchants will tell you where to buy coal grates.

Cost: $100 per ton. If you buy in bulk, perhaps $70 per ton.

Savings: Varies with your usage. Saves a few dollars to several hundred dollars per year.

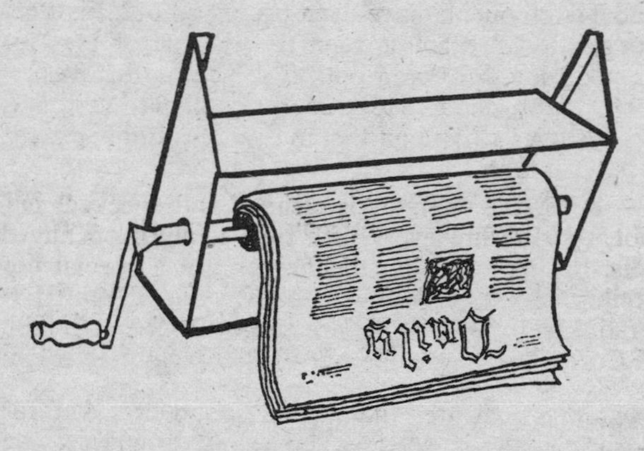

A paper-log roller.

Tip #41
Preventing Fires

Description: In 1978 there were 33,000 chimney fires in the U.S., causing $50 million in property damage. Poor woodburning practices eventually will cause a chimney fire in almost every home with a wood stove, experts believe.

Remedies: Before you burn wood at all—either in a fireplace or wood stove—have the flue checked by a reliable chimney contractor. Make sure there are no cracks in the mortar or other places where heat can escape. If metal ducts go though walls or ceilings, install insulating collars around them. Keep stoves away from walls and fabrics.

— Have the chimney thoroughly cleaned at least once a year. In the first year of heavy woodburning, have the ducts checked every two months for deposits of creosote, a flammable, tar-like substance in wood resins. Creosote is the culprit in many chimney and duct fires.

— After the chimney is cleaned and checked, make a hot, roaring fire every fifth time to burn off any deposits. Sprinkle salt on the fire or use a special commercial compound designed to keep flues clean. If you hear tinkling noises, they are pieces of creosote falling like Corn Flakes into the hot flames and safely burning off.

Where to find it: Chimney cleaners and contractors are listed in the Yellow Pages under "chimneys." Get at least 3 estimates for all work. Buy commercial anti-creosote preparations from stores that sell wood stoves.

Cost: $15 a year for the anti-creosote preparations.
$35–$50 a year for the chimney cleaning and inspection.

Savings: The value of your house, which won't burn down. Perhaps your life.

Epilogue: Buy a fire extinguisher. Keep it in a handy place.

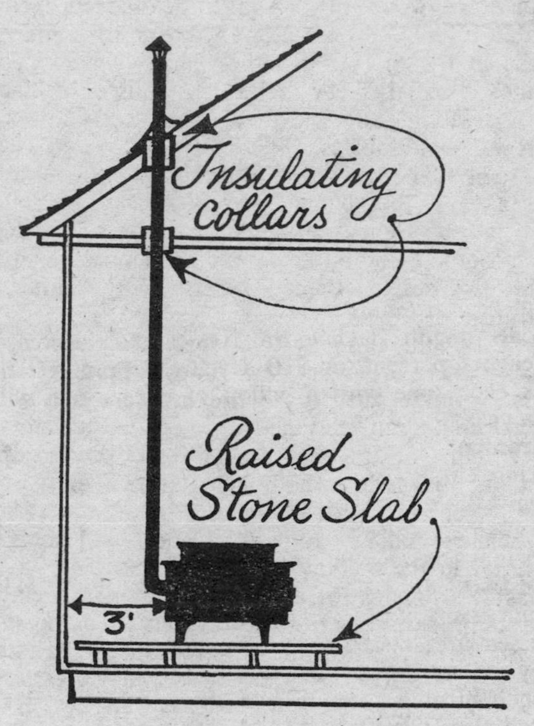

Correct installation of a duct.

7
Nickels and Dimes

Tip #42
Nickels and Dimes

Description: A small departure from our stated purpose of listing the biggest energy savers. Following are a few useful fine-tuning measures. Keep them in the back of your mind.

Car weight: Each extra 100 pounds decreases gas mileage 1 percent, or $10 a year. Topping off the gas tank each time with a gallon of excess fuel will cost $1 a year.

Ironing: An iron costs 6 cents per hour to operate. Taking clothes out of the dryer while they are still warm reduces ironing needs. If you do a week's ironing at one time, the iron has to warm up only once. This may save you $3–$5 a year.

Instant-On TV's: Picture and sound come on the instant the switch is flicked. The name is a misnomer. The sets are *always* on, even when the screen is blank. They are no longer sold, but millions still exist. Consider controlling the unit with a wall switch, or a switch placed on the TV cord. Saves $5–$10 a year.

Dust: Dusty lightbulbs emit 20 percent less light, meaning you might turn on another lamp. Dust off radiators for better heat transfer. Dust off motors and/or clean filters of refrigerators, air conditioners, electric clothes dryers, vacuum cleaners. Savings: a few dollars a year.

Light-colored roofs: Reduces summer heat gain in

cars and homes. Cuts air conditioning costs a couple percent.

Draining: Drain a pail or two of water and sediment from the spigot at the bottom of your hot water heater every six months. It keeps efficiency high and saves the equipment.

Drying: Drying several loads at once takes advantage of a warmed-up machine. Removing clothes to be ironed while still damp reduces drying time.

Cost: $0.

Savings: Perhaps $50 a year for all measures.

Epilogue: After you've saved all these nickels and dimes, don't waste them by leaving everything on when you go away on vacation. Unplug air conditioners, turn down heating and hot water thermostats, empty and unplug the refrigerator.

Instant-on, a misnomer.

8
Utility Companies and Banks

Tip #43
Utility Bills

Description: Customers often scrutinize restaurant tabs and complain loudly about 50-cent errors. But they pay thousands of dollars to utility companies without ever checking their bills.

— One Long Island, N.Y. woman overpaid by hundreds of dollars on her electric hot water for 18 years, because she had the wrong billing code. A friend found the error.

— You can use the utility company to reduce your energy bills. You just have to know what to ask.

Remedies:

— Check every utility bill—oil, gas, electric, propane, water, etc. Computers make mistakes. Read your meter. If you don't know how, call the utility's local customer service department and have someone explain the bills and meters to you. They must do it by state law. Many utilities have brochures.

— Ask the utility for an energy audit of your house. It usually costs $10 and will tell you the most cost-effective measures.

— Your utility company may have special reduced rates. One is an electric storage rate—a cheap nighttime electricity charge for heating hot water or recharging an electric car.

—A variation of this is time-of-use metering, which can cut a third off your bill. It is a voluntary rate—usually in an experimental program—that varies with the time of day: high during periods of peak demand, but very low late at night.

— Much of this information is in brochures enclosed with bills. Read them.

— If you don't get satisfaction from the utility, call the state agency that regulates utilities and complain. Usually, it is the public service commission or public utilities commission. Also, call newspapers: They may do a story.

Where to find it: Use phone book. You can obtain "How to Understand Your Utility Bill" from Consumer Information Center, U.S. Dept. of Energy, Pueblo, Colo. 81009.

Cost: $0.

Savings: As much as $200 per year.

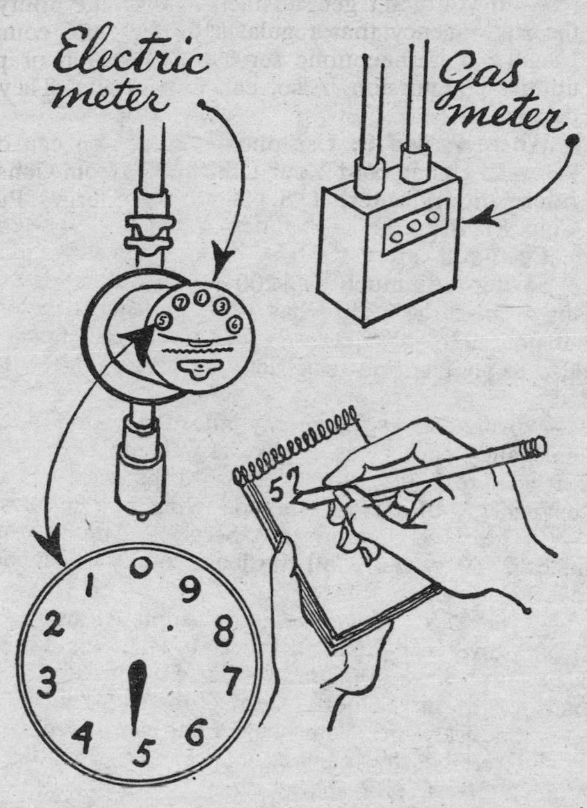

Computers make mistakes. Check your meter. Learn how to read your bill. You may discover a $100 error—or more.

Tip #44
How to Get Financing

Description: After you've picked the best energy conservation strategy, where will you get the money?

— While most devices are cheap, they still cost too much for many low-income and elderly residents.

— You might not have the ready cash for more expensive purchases, including contractor-installed insulation, storm windows or new furnaces.

Remedies: Call your local utility. Many arrange for financing from local banks for all sorts of improvements—from insulation to heat pumps. Often loans are at reduced interest rates—or no interest, as in the case of the Tennessee Valley Authority. This is especially true if the utility has done an energy audit of your house. In some cases, your loan payments can be added to your normal electricity bill in a budget-billing plan.

— Contact the state energy office for information on grants and loans. New York, for example, gives the elderly up to $200 in cash. It also dispenses $400 U.S. Department of Energy grants to low-income families.

— The U.S. Community Services Administration makes $350 grants. Call the local county or city anti-poverty or welfare office.

— The U.S. Farmer's Home Administration gives $1,500, five-year loans to rural low-income families. It also makes $5,000 grants (the "504" program) and loans for the rural elderly. Contact the local office.

— Shop around. Some banks and credit unions use low-interest energy loans as marketing tools. Also call your county or city energy office or clerk. Local officials sometimes work with banks to help residents finance energy-saving improvements.

Where to find it: Use the phone book.

Cost: $0; perhaps $200 in tax-deductable interest on loans.

Savings: Up to the entire cost of the device, plus the energy savings afterwards.

Tip #45
How to Get an Energy Tax Credit

Description: The federal government gives tax refunds to people who install energy conservation and alternate energy devices between April 19, 1977 and Jan. 1, 1986.

— Claim the credit for work done the year before. Save receipts. If the refund is more than you paid in taxes, you can carry over the excess to offset your next year's taxes.

— The state energy office may give additional tax credits. California, for example, gives a 55 percent tax credit on solar devices: more than the federal government's rebate.

Who qualifies: Both owners and renters, as long as it is the primary residence. If you move, you can claim the full credit again on your new residence.

Some of the devices that qualify for federal credits:

— Energy-Savers: Insulation, caulk, weatherstripping, storm doors and windows (including plastic), automatic thermostat setbacks and fuel-savings furnace modifications—flue dampers, new burners, electronic ignition replacing gas pilot lights.

— Alternate Energy: Solar heating, cooling and hot water equipment, wind-energy devices and geothermal energy equipment.

What does NOT qualify for federal credits:

— Energy-Savers: Primarily structural or decorative devices such as drapes, carpets, wood paneling, exterior siding.

— Alternate Energy: Solar swimming pool heaters, wood and coal stoves, heat pumps, fluorescent lights, new furnaces.

How much is the refund: For energy-saving devices, 15 percent of the first $2,000, or a maximum of $300. For alternate energy devices, 40 percent of the first $10,000, or a maximum of $4,000.

More information: Call the local U.S. Internal Revenue Service Center. Ask for publication 903 and form 5695, energy tax credits. Also contact your state energy office for state credits.

Form **5695**

Department of the Treasury
Internal Revenue Service

Energy Credits

▶ Attach to Form 1040.

Name(s) as shown on Form 1040

Residential Energy Credit Computation

This form may be worth thousands of dollars to you.

9
Buying and Building New Things

Tip #46
Air Conditioners

Description: One air conditioner can have twice the electric bills as an identically-sized unit. If you don't choose wisely, you can get burned while you're getting cooled.

— In hot climates with central units, it's also economical to reclaim the exhaust heat to warm water for washing.

Remedies: Choose an air conditioner with an EER —energy efficiency ratio—of at least 8.5. The EER selection is 4–12. The higher the number, the more efficient the unit—and the lower the fuel bills. The efficient units cost more to buy, but more-than-pay for themselves in lower operating costs.

— Don't buy an oversized air conditioner. It won't increase cooling. It will cycle on and off a lot, inefficiently, like an oversized furnace. Your dealer should have a form enabling you to estimate air conditioner size. A rule of thumb is 20 BTUs/hr. per square foot of floor space. Thus a 15-by-20 foot room (225 square feet) requires a 4,500 BTU/hr. air conditioner.

— In hot areas, it may pay to buy a heat exchanger that recycles air conditioner exhaust into the hot water system. The waste heat, removed from the dwelling's interior, could wipe out hot water bills during most of the cooling season. Units cost $400–$600 and save $150–$200 a year in hot water bills.

It's economical if you have at least 1,500 hours of annual central air conditioning and $300 in yearly hot water bills.

Where to buy it: Comparison shop. Use the Yellow Pages and call a number of dealers. If you can, get a list of all air conditioners and their EER's from the dealer or The Association of Home Appliance Manufacturers (see appendix). AHAM also has forms to estimate air conditioner size.

Cost/Benefit Analysis for a Room Air Conditioner:

— High efficiency unit: $400 to buy, $600 to run for 10 years. Total, $1,000.

— Low efficiency unit: $225 to buy, $1,200 to run for 10 years. Total, $1,425.

— High-efficiency unit saves $425. Resale value higher.

energy guide

Sears, Roebuck and Co. Model 76053

5,000 BTU per hour

(cooling capacity)

115 volts 795 watts 7.5 amperes

EER = 6.3

Energy Efficiency Ratio expressed in Btu per watt-hour

IMPORTANT . . . for units with the same cooling capacity, higher EER means:
Lower energy consumption
Lower cost to use!

For available 5,000 to 6,000 Btu per hour 115 volt window models the EER range is

EER 5.4 to EER 8.8

For information on cost of operation and selection of correct cooling capacity, ask your dealer for NBS Publication LC 1053 or write to National Bureau of Standards, 411.00, Washington, D.C. 20234

Data on this label for this unit certified by

Tested in accordance with

Look for this tag on air conditioners. Try to buy a unit with an EER of at least 8.5: The higher the number, the cheaper the unit is to operate.

Tip #47
Fences and Trees

Description: A simple, 5-foot-high picket fence can cut heating bills by a third. A line of trees can cut heating bills by 40 percent.

— By blocking the prevailing winds, you sharply cut the wind-chill factor, reducing the infiltration of cold air. Consider the lines of trees on windswept prairie farms.

— In one Nebraska test, a house with dense shrubs along one side had fuel bills 23 percent lower than an unprotected twin.

— In the summer, vegetation reduces cooling costs.

Method: Plant year-round trees or shrubs, or build fences along the windy side of your building. They should be at least 5 feet high. Install them 2½ windbreak heights away from the house. When planting, remember: Trees and shrubs will grow.

— On the sunny side, plant deciduous trees. In winter, their naked branches will let the warming sun through the windows. In summer, their leaves will shade the hot sun.

— Trees can lower the air temperature around your house 15 degrees. A 90-degree day becomes a 75-degree day, sharply cutting cooling costs. How? By drawing cool water from the ground and evaporating it through the leaves. A single maple can transpire 2,000–3,000 gallons a day. Lawns also provide cooling.

— Air conditioners in a fully-shaded house work only half as much as units in an unshaded house.

Cost: 10 evergreens, $600 installed; an average fence, $150 installed; shade trees, $50–$150 each, installed.

Savings: Heating—$200–$500 a year in windy areas.
Cooling—$150–$300 a year with a central system, $50 a year for two window units.

Where to buy it: Landscapers found in Yellow Pages. Fences at lumberyards and home improvement centers.

Epilogue: These items do more than save fuel. They

can raise the resale value of your property beyond their initial cost.

Sun warms house through tree branches.

Evergreens and fences reduce wind cutting heating bills.

Tip #48
Windows, Doors and Drapes

Description: When the outside temperature is zero, the inside of a single pane window is 17 degrees, collecting ice. But the inside of a double-pane window is 45 degrees, a triple-pane window, 55.

— During large temperature differences between inside and outside, you need a lot of barriers on windows. They include extra glass, drapes, shades and awnings.

Remedies: Buy drapes to use windows effectively. On sunny winter days, open drapes to let in the sun's warmth, cutting heating costs. On cold winter nights and hot summer days, close drapes to shield your living space. On summer nights, open drapes and windows, letting in breezes and cutting cooling bills. Drapes should be made of heavy material and touch the floor and both side walls.

— Add storm windows and doors. There are many different types: wood, aluminum, vertical, horizontal. Shop around. They cut heating and cooling bills 10–15 percent.

— Awnings or overhangs. In hot climates, consider them on the sunny side to cut cooling bills. In winter, the sun, which is lower in the sky, will cascade in under the obstructions.

— Roller shades or insulating shutters. May supplement or substitute for drapes. A matter of taste. Renters may prefer them to storms, because you can take them with you.

— All these items will more-than-pay for themselves. They add to a house's value and save energy. But they are initially expensive.

Where to buy it: Use the Yellow Pages. Roller shades are sold at such outlets as Sears, Montgomery Ward or J.C. Penney.

Cost: $60 each for storm windows, $200 for a storm door. $1,000 for an average house.

Prices for drapes, awnings and shades vary, from $100 to more than $1,000. Some people make drapes from corduroy.

Savings: $100–$150 a year for storm windows and

doors; $30 a year for drapes over double pane windows. Up to $150 a year for drapes, insulated shades and shutters on single pane windows.

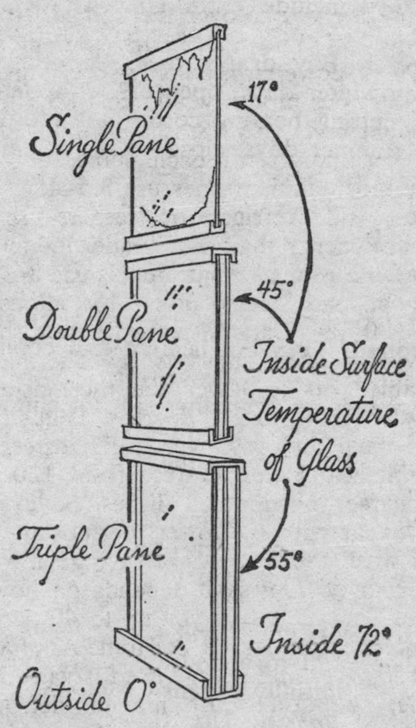

More glass, lower heating bills.

Tip #49
New Kitchen and Laundry Items

Description: Within the next 2 years, new major appliances will have government-required tags showing annual electricity use, so you can compare. Until then, your work will be harder.

Refrigerators: Try not to buy a new one unless you know its energy use. Otherwise, you may pay dearly: $2,300 more for power over the unit's 20-year life. A model using 85 kilowatt hours per month would cost $3,400 in power; an equally-sized unit using 140 kilowatt hours per month would cost $5,700.

— Efficient units use less than 5 kilowatt hours of electricity per month for each cubic foot of cooling space. The most efficient ones use 3.5–4.

— Side-by-side refrigerator freezers use up to 45 percent more energy than over/under models.

Dishwashers: Many recent models use 8–9 gallons of hot water per load instead of 14–15 gallons. This can save $30–$50 per year. Also, two new dishwashers—one by Kitchen Aid and one by Thermador/Waste King—enable you to lower the thermostat on your main hot water tank to 120 degrees. That's because the dishwaters have their own electric boosters that raise the incoming water temperature from 120 degrees to the 140 degrees needed for dishes. Savings: another $20–$40 a year with the lower thermostat.

Clothes washers: New water-saving units are available. The money they save depends on how hot your washes are. They will save you up to 40 percent.

More information: Assoc. of Home Appliance Manufacturers list water and energy use for various models.

Epilogue: We don't push manual defrost refrigerators, because they involve sacrifice. And, today's automatic defrosts are more efficient. But here are some economics for manuals, which use 25 less power than the most efficient automatics: Manual unit—$350 to buy, $2,750 to operate for 20 years. Automatic unit: $500 to buy, $3,400 to operate. Totals: Manual, $3,100; Automatic, $3,900.

Tip #50
Car and Tires

Description: At current monetary inflation rates, gasoline for a 20-miles-per-gallon car will cost $14,400 over its 10-year life. Gas for a 30-miles-per gallon car will cost $9,600 over the same period—$4,800 less.

— Gasoline is single largest outlay for a car. During an average car's life, the fuel to run it will cost far more than its original purchase price.

— Everyone knows gas mileage should be considered when buying a new car. But often, the process is not as organized or thorough as it should be. As a result, consumers don't get the most for their money.

Remedies:

— Get an up-to-date mileage guide issued by the U.S. Department of Energy (see Appendix). It will help you compare almost all cars sold in the U.S. If a single model has different-sized engines, buy the smallest one: It will use the least gas. The largest engine is only good for winning races on local streets.

— Use a calculator. Divide your average yearly mileage by the mileage rating of the model you're considering. This will give you your projected yearly gasoline use. Multiply this by the price of gas. Do it for five years, adding an extra 13 percent each year for inflation.

— Add maintenance, insurance and purchase price. Total all figures. Compare cost for each car.

— If you're indifferent to transmissions, consider a manual one: It will get 10 percent better gasoline mileage.

Cost: $0–$2,000 for more efficient model.

Savings: Up to $5,000—or more—in gasoline.

— Radial tires cut gas bills an average of 7 percent. They also last longer than standard tires. Buy four at once: You can't mix. The set will cost $200 more to buy, but will save $400 in gasoline and $160 in tires over the car's life.

Tip #51
Burners and Furnaces

Description: New technology has produced more efficient heating systems and components. Shop around. There are many types and combinations.

— With fossil fuel, get the smallest burner necessary to heat your house. Make sure the "steady-state" efficiency is at least 85 percent, and ideally above 90 percent.

— With electricity, find out whether your utility has an "off-peak" rate. If so, you can install equipment that stores heat at night, saving money with cheaper rates.

Oil: If you have a usable, tuned unit with less than 70 percent efficiency, buy a "retention head" burner. It will save 10–15 percent of your bills. You can get a 15 percent federal tax credit on it, too. Costs $250–$300; saves $100–$150 a year.

— For a new furnace, consider a small, highly-efficient "Blue Ray" unit. Costs about $1,800.

Gas: If you are converting from oil to gas, you may not need a whole new furnace, costing $1,500–$2,000. First consider a new, $500 burner using the same furnace.

— Gas is now much cheaper than oil, and will remain so at least until 1985. A conversion will pay for itself in 1–2 years.

Heat Pumps: If you are installing a central air conditioning system, consider an electric heat pump, which both heats and cools. It heats by extracting warmth from the outside. The colder it is outside, of course, the less efficient the heat pump. Ask how low the unit will go. In general, heat pumps use 40–60 percent less power to supply warmth than do conventional electric baseboards. A full system costs about $3,500.

— For 25–50 percent more, you can buy a heat pump that draws warmth from water: either in a tank connected to rooftop solar collectors, or from groundwater. In winter, the water is usually much warmer than the outside air, so these "water source" heat pumps are far more efficient than conventional ones. But ask about corrosion problems.

Where to buy it: Heating contractors and oil dealers. In all cases, comparison shop. Talk to previous customers.

Cost: Varies. Several hundred to several thousand dollars.

Savings: Can cut heating (and cooling) bills up to half.

Tip #52
Wood Stoves

Description: About 1 million wood stoves are sold each year—many to people who have not properly prepared. Result: Some stoves become fancy decorations, others use too much fuel, and yet others don't heat properly.

Remedies: First tighten up your house by the measures outlined in this book. Why buy an expensive wood stove and let the heat leak out uncaulked windows? A leaky house needs 75,000 BTUs of heat an hour; an efficient house, 25,000: meaning of a third of the wood and a smaller, cheaper stove. (Even an efficient house has enough leaks to prevent air pollution problems.)

— Secure a long-term wood supply *before* you buy a stove.

— Buy a stove at least 50 percent efficient. The joints should be fitted and cemented, castings precise, the inside lined with firebrick, the door sealed with asbestos. Ask the dealer for test results. Make sure the stove is approved by Underwriter's Laboratory. Talk to previous customers.

— Cast-iron stoves hold heat best and last longest. The walls should be at least a sixteenth-inch (16 gauge) thick. Exception: If you want quick, short fires, buy a steel stove. It has thinner walls. It is not meant for continuous operation.

— New masonry fireplaces are very inefficient (see Tip #38). Fireplace inserts are useful because you can use an existing space—but many have terrible efficiencies. Demand to see data by an *independent* laboratory. Talk to previous customers. The best use of a fireplace opening would be to put an efficient woodburning stove in it.

— Some insurance companies won't insure homes with woodburning stoves. Check before you buy.

Where to buy: Wood stove stores listed in the Yellow Pages or advertised in newspapers.

Cost: $400–$800 for high-quality stoves, depending on size. 3–6 cords of wood per stove each season.

Savings: The more you use it, the more you save.

You can make back the stove's cost in 1–2 seasons if you burn wood full time.

Epilogue: "Air tight stove" is a misnomer. The wood needs air to burn, although it is tightly controlled. Remember: Full-time woodburning involves work. Loading the stove. Banking the fire. Emptying the ashes. It's not easy.

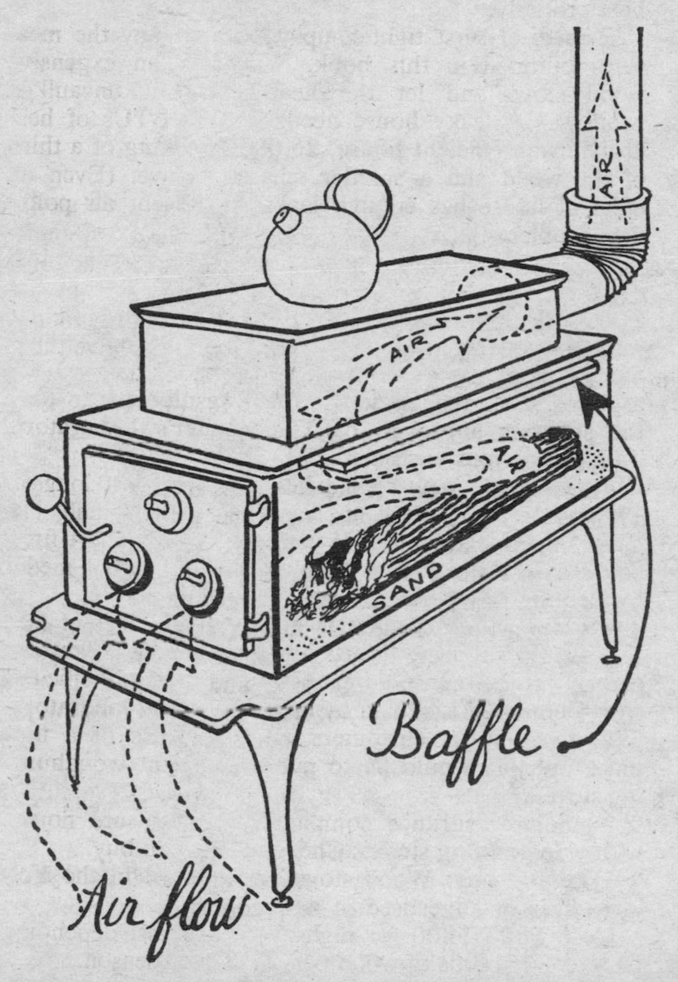

An efficient woodburning stove.

10

A Plan of Action

A Plan of Action

Description: You now understand the ways in which you can save hundreds and hundreds of dollars without sacrifice. The next thing is to organize your plan of attack.

(1) Collect information: Call the local utility and get an energy audit for $10. If there is a wait, have the utility send a form to do it yourself. The state energy office also may have do-it-yourself audit books. Write the U.S. Energy Department for more information. Visit the local library.

(2) Decide the order: Match the tips in this book to your house. Seal the cracks, stop the drips, insulate: Stop the obvious waste.

(3) Do one thing at a time: See what happens to your bills after each step. Don't do everything at once.

(4) Comparison shop: Use the Yellow Pages and newspaper ads. Do as much as you can yourself. Get estimates in writing for big jobs and make it clear to the contractor you know what's needed. Show him the pertinent information you've collected.

(5) Be there, if possible: Your presence—and questions—when contractors work will improve the quality of the job.

(6) Inspect the work before you pay

(7) Enjoy: You will reap the savings and satisfaction that comes from your new control over energy prices.

Table of Electric Appliance Use

Item	Wattage	Hours/year Operation	KWH/Yr.	Cost (6 cents/KWH)
AC (Room)	860	1,000	860	$51.60
AC (Central)	5,400	1,400	7,560	453.60
Blanket	150	1,000	150	9.00
Blender	380	40	15	.09
Broiler	1,440	75	108	6.48
Can Opener	100	3	0.3	.02
Clock	2.5	8,760	22	1.32
Clothes dryer	4,855	210	1,020	61.20
Clothes washer	510	200	102	6.12
Coffee maker	1,890	150	133.5	8.01
Corn popper	575	15	9	.54
Curling iron	40	50	1.6	.10
Dehumidifier	260	1,465	381	22.86
Dishwasher	1,200	300	360	21.60
Fan, attic	370	800	296	17.76
Fan, window	200	850	170	10.20
Freezer, frost-free (15 cubic feet)	440	4,000	1,760	105.60
Freezer, manual (15 cubic feet)	340	3,500	1,190	71.40
Frying pan (elec)	1,200	135	162	9.72
Hair dryer (hand-held)	1,000	42	42	2.52
Heat lamp (infrared)	250	50	12.5	.75
Heating pad	60	104	6.2	.37
Hot plate	1,260	75	94.5	5.67
Humidifier	175	943	165	9.90
Iron	1,100	104	115	6.90
Knife (elec)	95	8	0.8	.05
Light bulb	100	1,800	180	10.80

Table of Electric Appliance Use

Item	Wattage	Hours/year Operation	KWH/Yr.	Cost (6 cents/KWH)
Lighting (mixed wattage)	600	2,000	1,200	72.00
Microwave oven	1,450	130	188.5	11.31
Mixer	80	20	16	.96
Radio/record player	110	1,000	110	6.60
Range (w/oven)	12,000	100	1,200	72.00
Refrig/freezer (frostless, 14 cu. ft.)	615	2,975	1,830	109.80
Refrig/freezer (manual, 14 cu. ft.)	325	3,450	1,120	67.20
Roaster	1,425	72	103	6.18
Sewing machine	75	145	11	.66
Shaver	15	133	2	.12
Toothbrush (elec)	7	70	5	.30
TV color (tube)	300	2,200	660	39.60
TV color (solid state)	100	2,200	220	13.20
TV B/W (tube)	160	2,200	352	21.12
TV B/W (solid state)	55	2,200	121	7.26
Toaster	1,100	35	39	2.34
Vacuum cleaner	630	78	49	2.94

Conversion Tables

1 BTU (British Thermal Unit) — Amount of heat required to raise 1 pound of water (1 pint) 1 degree Fahrenheit

1 calorie — 4 BTUs.

1 kilowatt-hour — 3,412 BTU's. 1,000 watts of electricity (such as ten 100-watt bulbs) used for an hour.

1 gallon of gasoline — 125,000 BTUs.

1 gallon of fuel oil — 140,000 BTUs.

1 barrel of oil	—	42 gallons.
1 ton of coal	—	26 million BTUs.
1 cubic foot of natural gas	—	1,000 BTUs.
1 therm of natural gas (100 cubic feet)	—	100,000 BTUs.

BTU's, TNT, Polyester Suits

A match	1 BTU	
An apple	400 BTU	
Making a cup of tea	500 BTU	
A stick of dynamite	2,000 BTU	
A large piece of chocolate layer cake	2,000 BTU	
A loaf of bread	5,100 BTU	
A pound of wood	6,000 BTU	
100 hours of television	28,000 BTU	
A gallon of gasoline	125,000 BTU	
Cooking with a gas range for 20 days	1,000,000 BTU	(1 million)
Food for one person for a year	3,500,000 BTU	(3.5 million)
Making a polyester suit	20,000,000 BTU	(20 million)
Heating a typical house in St. Louis for a year	90,000,000 BTU	(90 million)
Apollo 17 to the moon	5,600,000,000 BTU	(5.6 billion)
Atomic bomb dropped on Hiroshima	80,000,000,000 BTU	(80 billion)
A thousand transatlantic jet flights	2,500,000,000,000 BTU	(2.5 trillion)
Energy used in one year in Oklahoma	1,000,000,000,000,000 BTU	(1 quadrillion)
Energy used in one year by 30 African countries	1,000,000,000,000,000 BTU	(1 quadrillion)
Energy needed to boil Lake Erie	1,600,000,000,000,000 BTU	(1.6 quadrillion)
Energy used by U.S. in 1975	75,000,000,000,000,000 BTU	(75 quadrillion)
Energy used by entire world in 1975	225,000,000,000,000,000 BTU	(225 quadrillion)

Appendices

Sources for Further Information

(1) Consumer Information Center, U.S. Department of Energy, Pueblo, Col. 81009. Publishes dozens of free brochures on energy conservation.

(2) Association of Home Appliance Manufacturers, 20 North Wacker Drive, Chicago, Ill. 60606. Brochures on buying major appliances, list of air conditioners and their efficiencies, tips on saving energy wth appliances.

(3) Fuel Economy Distribution, Technical Information Center, U.S. Department of Energy, P.O. Box 62, Oak Ridge, Tenn. 37830. Distributes automobile mileage ratings.

(4) National Technical Information Service (NTIS), Springfield, Va. 22161. The central source that sells government-sponsored research. It has thousands of documents describing almost every imaginable type of energy. Has studies on heating systems, cars, and other energy subjects of concern to consumers. Consult index at local library for stock numbers of appropriate reports. Or call 202–724–3382; 703–557–4650.

(5) National Weather Bureau, Federal Building, Asheville, N.C. 28801. Its booklet "Climatic Data," lists amount of sunshine, temperature, humidity, wind speeds, cloudiness, precipitation and heating and cooling degree days for about 300 locations around the country.

(6) Institute of Real Estate Management, 430 North Michigan Ave., Chicago, Ill. 60611 and Community

Housing Improvement Program (CHIP), 575 West End Ave., New York, N.Y. 10024. Two organizations of apartment owners, they have energy programs designed to show landlords how to cut fuel bills. Since such measures would reduce the rent, you as a tenant might put your building management in touch with the two.

(7) Wood Energy Research Corp., P.O. Box 800, Camden, Maine 04843. 207–236–8575: Publishes booklet on starting a wood-buying collective and has up-to-date information on the benefits—and pitfalls—of wood-burning.

ABOUT THE AUTHOR

STUART DIAMOND covers energy and environment for *Newsday*, the Long Island newspaper, and is a contributing editor for *Omni*, the science magazine. His first book, *It's in Your Power*, dealt with energy survival and was coauthored with Paul S. Lorris in 1978. Since then, Diamond has written cover stories on energy and environment for magazines such as *Reader's Digest*, *Family Circle* and *Family Health*. Diamond also has appeared on ABC TV's *Good Morning America*, NBC's *Today Show*, *CBS News* and other radio and TV programs to discuss energy problems and solutions. His numerous awards include citations from three journalism societies for stories on the nuclear accident at Three Mile Island and an award from the business school at Dartmouth College for a series on the growing problems of garbage. In 1979, a newspaper series he coauthored on the gasoline crisis won the prestigious Polk Award for national reporting. Stuart Diamond lives in an old Victorian house that he is trying to renovate, caulk and weatherstrip in Northport, N.Y.

ABOUT THE ILLUSTRATOR

PATRICIA WINDROW, the illustrator, was born in London, England and has been a practicing artist for four decades. She has done commercial illustration for major American magazines and manufacturers, and has won numerous prizes for her fine arts painting. In 1980, she was included in *Who's Who in American Art*. Ms. Windrow has illustrated six books, including *My Best Friends Are Dinosaurs* and *The Three Village Guide Book*. Patricia Windrow owns and operates the Phantom House Art Gallery in Setauket, Long Island, N.Y.